The Mystery of Rationality

Gérald Bronner · Francesco Di Iorio
Editors

The Mystery of Rationality

Mind, Beliefs and the Social Sciences

 Springer

Editors
Gérald Bronner
Paris Diderot University
Paris, France

Francesco Di Iorio
Nankai University
Tianjin, China

and

French Academy of Technologies
Paris, France

and

French National Academy of Medicine
Paris, France

ISBN 978-3-319-94026-7   ISBN 978-3-319-94028-1   (eBook)
https://doi.org/10.1007/978-3-319-94028-1

Library of Congress Control Number: 2018945437

Printed on acid-free paper

This Springer imprint is published by the registered company Springer International Publishing AG part of Springer Nature
The registered company address is: Gewerbestrasse 11, 6330 Cham, Switzerland

To the memory of Raymond Boudon

Contents

1 Introduction: Rationality as an Enigmatic Concept   1
Gérald Bronner and Francesco Di Iorio

2 Our Agenda and Its Rationality   7
Joseph Agassi

**3 Intentional, Unintentional and Sub-intentional Aspects of Social
Mechanisms and Rationality: The Example of Commitments
in Political Life**   17
Alban Bouvier

**4 On the Explanation of Human Action: "Good Reasons", Critical
Rationalism and Argumentation Theory**   37
Enzo Di Nuoscio

5 Rationality, Irrationality, Realism and the Good   53
Paul Dumouchel

**6 First Generation Behavioral Economists on Rationality,
and Its Limits**   67
Roger Frantz

7 Embodied Rationality   83
Shaun Gallagher

8 Rational Choice Explained and Defended   95
Herbert Gintis

**9 Rationality and Irrationality Revisited or Intellectualism
Vindicated or How Stands the Problem of the Rationality
of Magic?**   115
Ian Jarvie

10 Rational Life Plans?   131
Daniel Little

11 **Dynamics of Rationality and Dynamics of Emotions**............ 147
    Pierre Livet

12 **Pathologizing Ideology, Epistemic Modesty and Instrumental
    Rationality** 165
    Leslie Marsh

13 **Do the Social Sciences Need the Concept of "Rationality"?
    Notes on the Obsession with a Concept**.................... 191
    Karl-Dieter Opp

14 **Rationality of the Individual and Rationality of the System:
    A Critical Examination of the Economic Calculation Problem
    Over Socialism** 219
    Ennio E. Piano and Peter J. Boettke

15 **Rationality and Interpretation in the Study
    of Social Interaction** 239
    Emmanuel Picavet

Contributors

Joseph Agassi Tel Aviv University, Tel Aviv-Yafo, Israel; York University, Toronto, Canada

Peter J. Boettke George Mason University, Fairfax, USA

Alban Bouvier Aix-Marseille University, Marseille, France; Institut Jean Nicod (CNRS/ENS/EHESS), Paris, France

Gérald Bronner Paris Diderot University, Paris, France; French Academy of Technologies, Paris, France; French National Academy of Medicine, Paris, France

Francesco Di Iorio Nankai University, Tianjin, China

Enzo Di Nuoscio University of Molise, Campobasso, Italy; LUISS Guido Carli University, Rome, Italy

Paul Dumouchel Graduate School of Core Ethics and Frontier Sciences, Ritsumeikan University, Kyoto, Japan

Roger Frantz San Diego State University, San Diego, CA, USA

Shaun Gallagher University of Memphis, Memphis, USA; University of Wollongong, Wollongong, Australia

Herbert Gintis Santa Fe Institute, Santa Fe, NM, USA

Ian Jarvie York University, Toronto, Canada

Daniel Little University of Michigan-Dearborn, Dearborn, USA

Pierre Livet CEPERC, Aix Marseille University, Aix-En-Provence, France

Leslie Marsh The University of British Columbia, Vancouver, Canada

Karl-Dieter Opp University of Leipzig, Leipzig, Germany; University of Washington, Seattle, USA

Ennio E. Piano George Mason University, Fairfax, USA

Emmanuel Picavet Université Paris 1 Panthéon-Sorbonne, Paris, France

Chapter 1
Introduction: Rationality as an Enigmatic Concept

Gérald Bronner and Francesco Di Iorio

Abstract It is evident that it is impossible to provide a commonly accepted defi-nition of rationality, and that there is a lack of agreement on the meaning of the concept. As a consequence, it can be said that there is a 'mystery of rationality'. What is it to be rational? The disagreements concerning the meaning of rationality can be related to (often intermingled) debates on six well-known dichotomies: (i) normative versus descriptive; (ii) instrumental versus non-instrumental; (iii) Cartesian versus non-Cartesian; (iv) tacit versus explicit; (v) explanation ver-sus interpretation; and (vi); intended versus unintended.

The analysis of the concept of rationality, which is central to various research fields, is a leitmotiv in the history of the social sciences and has involved endless disputes. It has been argued that rationality can be recognized and understood, but cannot be defined (e.g. Boudon 1995). According to this view, it could be said about rationality what Saint Augustine ([398]2009, p. 244) famously said about time: "What, then, is time? If no one asks me, I know what it is. If I wish to explain it to him who asks me, I do not know". The indefinability thesis is by no means universally shared. However, it is evident that it is impossible to provide a commonly accepted defi-nition of rationality, and that there is a lack of agreement on the meaning of the

G. Bronner
Paris Diderot University, Paris, France
e-mail: gerald.bronner@univ-paris-diderot.fr

G. Bronner
French Academy of Technologies, Paris, France

G. Bronner
French National Academy of Medicine, Paris, France

F. Di Iorio
Nankai University, Tianjin, China

concept. As a consequence, it can be said that there is a 'mystery of rationality'. What is it to be rational? The disagreements concerning the meaning of rationality can be related to (often intermingled) debates on six well-known dichotomies: (i) normative versus descriptive; (ii) instrumental versus non-instrumental; (iii) Cartesian versus non-Cartesian; (iv) tacit versus explicit; (v) explanation versus interpretation; and (vi); intended versus unintended.

Normative/descriptive dichotomy. Rationality was originally understood as the analysis of the rules of correct meaning. Many thinkers, such as, for example, Aristotle, Cicero, Malebranche, Descartes, Hume, Condorcet and Kant, have tried to define those rules and warned against false evidence and logical fallacies. The concept of rationality was thus for long regarded as essentially normative. According to this interpretation, this concept provides a criterion of rationality, i.e. a principle of objective optimization that allows one to establish whether or not an action is rational. This normative theory of rationality, which is still very popular today especially in economics, but also in broad sectors of sociology and psychology (regarding psychology, consider, for example, Amos Tversky's and Daniel Kahneman's prospect theory), is challenged by a descriptive theory of rationality. According to the latter, rationality is not an objective principle, but a simply explanatory assumption that can be applied also to actions that fail to fulfill the objective optimization standard because they are based on false and mistaken beliefs. On this view, rationality must be conceived in wider terms, i.e. as subjective, bounded and situated. In addition, false and mistaken beliefs must be regarded as the product of the normal ways of functioning of the human mind rather than of illogical tendencies. The descriptive notion of rationality is rooted in argumentation theory, hermeneutics and interpretative sociology, and it is supported also by heterodox economists such as Ludwig von Mises and Herbert A. Simon.[1]

Instrumental/non-instrumental dichotomy. According to a view widespread in economics, but also in other fields, rationality is purely instrumental, which means that it must be defined in terms of appropriate choice of means to achieve a goal. There are both normative theories of instrumental rationality (e.g. Gary Becker) and non-normative theories of instrumental rationality (e.g. Simon), but all share the assumption that, while the choice of means can be explained in rational terms, the choice of values, understood as the ends of human action, usually cannot and must be explained either in terms of illogical tendencies or in terms of socio-cultural determinism. By contrast, the non-purely instrumental theory of rationality (e.g. Raymond Boudon), which is rooted in the hermeneutic tradition and supported by major sectors of interpretative sociology, assumes that even the choice of values is rational, and that rationality is not exclusively instrumental. According to this wider

---

[1]According to some supporters of the descriptive theory of rationality such as Ludwig von Mises, even akratic actions, i.e. actions characterized by weakness of will, must be regarded as rational because, even if the agent will later regret having carried them out, akratic actions attempt to "remove a certain uneasiness" (1998 p. 15) in the way that the agent considers, when he/she acts, the action most appropriate given his/her subjective knowledge.

theory of rationality, the endorsement of values is rational not in the sense of a Cartesian and demonstrative rationality, but rather in that of an argumentative rationality based on rhetorical and intuitive skills, i.e. on what Pascal called *spirit of finesse* as opposed to the *spirit of geometry*. Supporters of this wider theory of rationality argue that the ability to perform value-judgments is part of human intelligence.

Cartesian/non-Cartesian dichotomy. According to the Cartesian tradition, only actions based on clear demonstrative arguments (e.g. $2 + 2 = 4$) can be regarded as rational, whereas actions whose meaning cannot be explained in explicit, precise and clear manner cannot. This Cartesian theory of rationality, which is normative, is exemplified by the mind/computer analogy developed in cognitive science. This view on rationality is challenged by a non-Cartesian approach. According to the latter, even actions that do not stem from precise demonstrative reasoning such as value-judgments based on vague arguments must be regarded as the product of human intelligence and its interpretative skills, which are partly tacit. This second theory of rationality is rooted in the argumentation theory and is exemplified by the so-called 'new rhetoric' (e.g. Chaïm Perelman; Lucille Olbrechts-Tyteca).

Using the first three dichotomies outlined above, it is possible to argue that the dominant model of rationality in economics is different from the model of rationality used in sociology (see Sen 1993). This is because the first is instrumental, Cartesian (axiomatized) and normative, while the second, which is more closely linked to the hermeneutic tradition, is quite often non-Cartesian, descriptive and not purely instrumental (see Boudon 1993).[2] As pointed out by Lévy-Garboua (1981, p. 30), sociologists tend to conceive rationality in terms of simple intentionality, while economists tend to conceive rationality in terms of effectiveness, which means that economists tend to use a notion of rationality narrower than the one employed by sociologists.

Tacit/explicit dichotomy. The connection between intentionality and rationality, which is acknowledged by the majority of economists and sociologists, is far from being unproblematic. It is partly rejected by neurosciences, which usually define intentionality in terms of "conscious will" as opposed to the sub-intentional mental activities, i.e. the so-called tacit skills. Neurosciences maintain that, since conscious skills are anatomically intermingled with tacit skills and cannot be clearly distinguished from the latter, rationality is linked to our overall mental activity, which cannot be reduced to our consciousness (see Berthoz 2003; Cleeremans 2003; Naccache 2006; see also: Lachaux 2013; Brass and Haggard 2007). Neurosciences have shown, for example, the importance of nonverbal reasoning, i.e. of a kind or

---

[2]This point is stressed not only by sociologists, who often criticize the model of rationality used in economics, but also by many economists (e.g. Sen 1977; Vandberg 1994; Ben-Ner and Putterman 1998). However, this difference should be not regarded as excessively radical. As argued by Wolfesperger (2001), there are a number of works in econometrics that consider ethics and social prestige as important explanatory factors. See also Gätchter and Ferhr (1999).

reasoning that is prior to our verbalized and conscious mental activities. This analysis of the relationship between intentionality and rationality developed by neurosciences is of great significance from a philosophical viewpoint because it induces, among other things, a rethinking of the ancient problem of the freedom of the will. This is because the findings of neurosciences entail that it is problematic to consider the conscious will as the foundation of the decision-making process.

Explanation/interpretation dichotomy. According to some authors, such as Wilhelm Dilthey and Benedetto Croce, rationality, understood as an explanatory assumption, is inconsistent with the principle of causality. This is because explaining the motivation of an action requires understanding its meaning by 're-experiencing' the thoughts and feelings of the agents. This empathic view of understanding, which supports a radical difference between the methodology of the natural sciences and that of the social sciences, is rejected by Carl Gustav Hempel and other authors. The latter argue that explanations in terms of rationality are, like explanations in natural sciences, causal explanations based on the deductive-nomological model. According to this anti-dualist epistemology, understanding the reasons of action presupposes the use of the causality principle and the determination of a cause-effect relationship through covering laws. Hempel and the other supporters of the unity of science stress that the covering laws used in the social sciences are usually commonsense laws about human behavior that are, like some laws employed in natural sciences, probabilistic rather than deterministic.

Intended/unintended dichotomy. Another problematic aspect of the concept of rationality consists in the tendency to overestimate its range of applicability that is rooted in the old religious interpretations of social phenomena in animistic terms. The dangers of such a tendency, which still exists in some sectors of the social and political thought, are particularly evident in the case of conspiracy theories. The above-mentioned tendency is well-known by psychologists, who use different expressions to refer to it: for example, "illusion of external agency" (Gilbert et al. 2000), "hyperactive agency detection" (Tempel and Alcock 2015), and "biased attributions of intentionality" (Brotherton and French 2015). Authors such as Max Weber, Carl Menger, Friedrich Hayek and Karl Popper, following the Scottish Enlightenment thinkers, cautioned against the simplistic analysis of social phenomena in terms of intentionality. They argued that quite often, because of their complexity, these phenomena cannot be interpreted as intended outcomes. On this view, the understanding of the social world depends on the study of both the individual's intentions, which are related to rational evaluations, and the emergent unintended consequences, which are, to use Adam Ferguson's words, "the result of human action, but not the execution of any human design" (Ferguson 1767–1996, p. 119).

The debates related to the six above-mentioned dichotomies inform the fifteen chapters of this book, which analyze issues related to rationality from different standpoints. It is the hope of both the editors and the authors of this work that it may contribute usefully to the above-mentioned debates and shed some light on the mystery of rationality.

References

Augustine, Saint. (2009). *Confessions*. Oxford: Oxford University Press.

Ben-Ner, A., & Putterman, L. (1998). Values and institution in economic analysis. In A. Ben-Ner & L. Putterman (Eds.), *Economic, values, and organization*. Cambridge: Cambridge University Press.

Berthoz, A. (2003). *La décision*. Paris: Odile Jacob.

Boudon, R. (1993). Toward a synthetic theory of rationality. *International Studies in the Philosophy of Science, 7*(1), 5–19.

Boudon, R. (1995). *Le juste et le vrai*. Paris: Fayard.

Brass, M., & Haggard, P. (2007). To do or not to do: the neural signature of self-control. *The Journal of Neuroscience, 22–27*(34), 9141–9145.

Brotherton, R., & French, C. C. (2015). Intention seekers: Conspiracist ideation and biased attributions of intentionality. *PLoS ONE, 10*, e0124125. https://doi.org/10.1371/journal.pone.0124125.

Cleremans, A. (2003). *The Unity of Consciousness: Blinding, Integration, and Dissociation*. Oxford: Oxford University Press.

Ferguson, A. (1767–1996). *An Essay on the History of Civil Society*. Cambridge: Cambridge University Press.

Gätchter, S., & Ferhr, E. (1999). Social norms as a social exchange. *Journal of Economic behavior and Organisation, 39*, 341–369.

Gilbert, D. T., Brown, R. P., Pinel, E. C., & Wilson, T. D. (2000). The Illusion of External Agency (PDF). *Journal of Personality and Social Psychology, 79*(5), 690–700. https://doi.org/10.1037/0022-3514.79.5.690. (PMID 11079235. Archived from the original on 2016-03-04).

Lachaux, J.-P. (2013). *Le cerveau attentif*. Paris: Paris, Odile Jacob.

Lévy-Garboua L. (1981, 30), «L'économique et le rationnel». *L'Année sociologique*, XXXI, pp. 19–46.

Naccache, L. (2006). *Le nouvel inconscient*. Christophe Colomb des neurosciences, Paris, Odile Jacob: Freud.

Sen, A. (1977). Rational fools: A critique of the behavioral foundations of economic theory. *Philosophy & Public Affairs, 6*, 317–344.

Tempel, J., & Alcock, J. E. (2015). Relationships between conspiracy mentality, hyperactive agency detection, and schizotypy: Supernatural forces at work? *Personality and Individual Differences, 82*, 136–141.

Vandberg, V. J. (1994). *Rules and choice in economics*. London: Routledge.

Wolfesperger A. (2001). «La modélisation économique de la rationalité axiologique. Des sentiments moraux aux mécanismes sociaux de la moralité». In Boudon, Demeulenaere et Viale (eds.) *L'explication des normes sociales*, Paris: PUF.

Chapter 2
Our Agenda and Its Rationality

Joseph Agassi

Abstract The rationality of conduct is its adequacy to the goal at which it is intended to aim. It seems obvious, until one attempts to apply it. Many cautionary tales report investments of effort to reach goals that, it turns out then, are the wrong goals, mistakenly deemed the right ones. How do people choose their goals? How is it possible to reduce the likelihood of error in the choice of one's goals?

Consider actions more important than reading books: research. My teacher, Karl Popper, was a great philosopher who devoted most of his time to research. Still, he had to choose the items for research. As it happened, he devoted much of his working life to the refutation of the currently received opinion on induction—the opinion that the choice of hypotheses is justified by their empirical support as this support follows the calculus of probability. He refuted it to his own satisfaction in 1935, and produced brilliant counter-examples to it in 1956. In his terrific review of *The Logic of Scientific Discovery* of 1959 (*Logik der Forschung*, 1935), Mario Bunge says, Popper has crushed the opinions that he was criticizing. His philosophy differs from that of Popper, but throughout his long, rich and respected career, he considered Popper's criticism deadly. To the best of my knowledge, Popper did not know of Bunge's review of his great book, perhaps since he did not read Spanish. (The review appeared in *Ciencia e Investigacion*, 15, 1959.) It is a pity.

Throughout his career, Popper argued against the received opinion on induction in the hope to alter it. In the preface to the 1984 edition of his *Logik der Forschung* he said of the penultimate appendix to it (Appendix *19) that it puts an end to Aristotelian induction and forces us back to the Socratic method of conjectures and refutations. (He used the expression "we must".) The appendix ends with a complaint about the oversight of his earlier refutations. This one is overlooked too. His last appendix (Appendix *20) published in his last year, 1994, offers yet a newer

J. Agassi (✉)
Tel Aviv University, Tel Aviv-Yafo, Israel
e-mail: agass@post.tau.ac.il

J. Agassi
York University, Toronto, Canada

© Springer International Publishing AG, part of Springer Nature 2018
G. Bronner and F. Di Iorio (eds.), *The Mystery of Rationality*,
https://doi.org/10.1007/978-3-319-94028-1_2

7

refutation, from his joint paper with David Miller. Most philosophers still cling to inductivism: it remains the received opinion.

Question: is the continuation of this battle incumbent on us—on those who side with Popper against the received opinion that scientific theories are inductively justified? I think not. In my view, Popper made a serious effort when he deemed his opponents rationalists and more so when he deemed it worthwhile to fight for the souls of rationalists by using the force of arguments to make them change their minds (Plato, *Gorgias* 457–8). Quite possibly, as he taught us to say, his view is true and mine is mistaken. I still say, as a point of order, that my disagreement with him has priority over his disagreement with the philosophical community. Is this true? How do we decide such matters? How do we decide our agendas? This is the problem of the present discussion. (And if you ask, who are we? Then the Socratic answer is, you and I.)

A classical paper by arch-physicist Erwin Schrödinger, "Are There Quantum Jumps?" (*The British Journal for the Philosophy of science* 3, 1952, 109–123) led arch-physicist Max Born to express anger at the very fact that Schrödinger had even re-raised the question ("The interpretation of quantum mechanics." *The British Journal for the Philosophy of Science* 4, 1953, 95–106). It was not agenda, he claimed, and for a good reason. Schrödinger was cognizant of the fact that he was deviating from established public agenda of physics. He opened his paper expressing a wish to take distance from the situation then current. To that end, he began that famous paper of his with the Copernican hypothesis.

Since then, leading historian of science Alexandre Koyré has discussed the question, how did the astronomical community learn about Copernicus, how did his ideas enter the astronomical agenda. He succeeded in putting this question on the agenda of historians of science, which is no small matter. I do not know what stir the Copernican hypothesis made in the mid sixteenth century; by the turn of the century, however, it hardly made an impression. Its most famous advocate was flamboyant Giordano Bruno whom the Church of Rome burned at the stake as a heretic in the year 1600. He was the last martyr of science, but most of the literature does not refer to him as a scientist at all. It considers Galileo and Kepler Copernicus' legitimate heirs. Kepler heard about Copernicus from Michael Maestlin, whom he knew personally. Galileo learned it otherwise. Christian Wursteisen or Urstisius, a Swiss itinerant lecturer then—one who made a living by traveling and lecturing, I do not know how exactly—came to Florence to lecture in some academy, these are the cryptic words that old Galileo used to tell the story (first *Dialogue*, Stillman Drake's translation 1970, p. 128), where young Galileo was hanging about as a college drop-out: he said he had no time to waste in college. He did not hear the lecture, as he deemed the Copernican hypothesis cranky. People who did go to the lecture told him about it and kindled his interest.

This is how items not on the public agenda enter: by default. Galileo's telescope put the Copernican hypothesis on the public agenda. Kepler's laws made this irreversible. Possibly, if not the one event then the other would have done it; this is hard to judge.

Let me mention another striking example: the way Maxwell learned about Faraday's theories. I have narrated in detail the strange chain of events that led him to study Faraday's ideas although they were non-agenda (*Faraday as a Natural Philosopher*, 1971): he heard about them indirectly from a cousin of Faraday who was a temporary lecturer in a university where the professor was sick. I have also mentioned evidence that in the twentieth century on the average worthy scientific ideas that emerged from non-Establishment of science received public attention two decades later than innovations from Establishment sources ("Cultural Lag in Science", in my *Science and Society: Studies in the Sociology of Science*, 1981). Stories that we read in history books suggest that a new idea naturally receives attention. These stories are too smooth and they blind us to the need to discuss ways to improve matters, especially the way to make good ideas publicly accessible as early as possible. My aim here is to put this on the agenda. Thus far, I have failed. (A leading reviewer of my book on Faraday said, it has nothing new.)

How do people fix their agendas? The question has an obvious answer. Decisions on agendas, both private and public, depend on orders of priority of questions to discuss and the efforts that each discussion demands. Moralists, religious or Socratic, demand that we mend our agendas and attend to our salvation first. This discussion is important, since self-deception rests on small errors that take important critical discussions off agendas—personal or also public, especially in politics. This is particularly so: the question of private agendas, though central to Plato's *The Apology of Socrates*, was banished from public discussions since the Age of Reason. Public discussions of the public agenda are rare too. Francis Bacon said, the wish to advance humanity should lead not to political activity but to research, since scientific advancement will cause improvement better—even when judged politically. How is the agenda of research to be determined? Immanuel Kant explained the very existence of a public agenda of science by his theory known as the architectonic unity of pure reason and as the unity of the apperception. It is a neglected part of his philosophy: philosophers ignore his idea and there is no alternative to it.

Fixing public agendas follows public rules that differ in different times and places. They differ also as to quality: some rules are better than others, and so they are comparable. The comparison is relative to the aims of the group that endorses the rules for fixing agendas.

The group that is topic of this discussion is the commonwealth of learning; not only because of my own interest personally, but also because of my view of this agenda as both very important and relatively easy to discuss. True, there are more important items (such as world peace), but they are much more difficult to discuss. As the commonwealth of learning is self-selected and open, it has no steering committee. Members are at liberty to choose their agendas. This enables me to discuss the agenda without asking for permission, and likewise it allows you to ignore my discussion. In the period following World War II, new factors entered the situation: almost all members of the commonwealth of learning are now academics, academies follow established rules of agenda, and at times they pertain to admission and grant giving and thus to research. As the commonwealth of learning is still varied, this may be a marginal matter. Generally, no research is either purely private or purely public, since the motive of research is in part private and the efforts put into research and into the publication of its results are in part public. At least the very presence of the information highways has greatly facilitated the access of private scholars to public arenas. This too is not a part of my present discussion—I have discussed it elsewhere ("Externalism", reissued in my *Science and Society, ibid.*).

My interest in agendas began when I was a student and heard in a science class the observation that around the year 1900 there were three cardinal problems in physics. I have later discussed one of them in some detail (*Radiation Theory and the Quantum Revolution*, 1993). What intrigued me at the time was the question, how do diverse researchers in one field reach a consensus on such matters? In my entire career, I found only two philosophers who discussed this matter, Sir Francis Bacon and Michael Polanyi. They are utterly unhelpful. Bacon said, follow every lead. This is obviously impossible. He knew this and said, we need teams for that. This raises the problem of coordination and so of agendas (*The New Atlantis*, 1627). He did not discuss that. Polanyi said, the leaders of a field orchestrate research in it (*Personal Knowledge*, 1958). They decide which leads to follow. He refused to discuss the question, how they come to such decisions. He said, this is a matter of skill, of personal knowledge, and it is impossible to describe skills—scientific and artistic alike. He was right to point out that no description of a skill is exhaustible, but then anyway no description is. Some descriptions of skills exist—of skills in politics and in business and in other areas of public affairs and even in the arts. He thought poorly of them. Still, we do describe the skills of orchestrating—a symphony or research—and the partial success that these ventures obtained. It is possible to start with efforts to describe the public decisions about priority of scientific problems.

Comments of friends have baffled me: will my audience care for discussions about fixing agendas? Will they consider it a philosophic matter? I do not know. True, some may say, as it is not an item on the publicly acknowledged philosophical agenda, it does not concern me. This is unobjectionable, of course. I have encountered such comments delivered with hostility too. This is regrettable. We have an important question here, be it philosophical or not, popular or not. Presumably, even those who limit their interest to the current agenda of philosophy may find a problem attractive and interesting enough to pay attention to it for a brief while regardless of the publicly acknowledged agenda of some academic discipline.

The problem of the order of priority of problems accompanies all concerns. There is too little discussion of agendas anywhere in the literature, even in political science. A former member of the British parliament once told me that he had made great efforts in putting on the parliament's agenda a proposal for a bill without noticing that it had no chance of getting to the top of the agenda to gain discussion time before elections to a new parliament erased the agenda and set it anew. Only the very few items on the top of any agenda ever gain attention. This is why there are different organizations and each has different steering committees that are in charge of their agendas. At times, of course, a minor body manages to alter the agenda of a more significant body. Thus, the ecological movement tries hard to put concerns for the ecosystem on the agendas of powerful national and international bodies. They fail because they do not know how to do their own homework of discussing their own agenda: they find too many important items and they do not know how to order their priorities. No one has the right to blame them, though, as there is no known procedure for this kind of activities, since too many vital items crowd the agenda.

That the agenda is of the utmost importance is no news. Nor is it news that it is advisable to avoid putting excessive constraints on the agenda and it is advisable to avoid leaving it too open. Running a semblance of democracy is easy: despots achieve such fraudulent charade by instituting ineffective parliaments without democratization. They achieve this with ease simply by not allowing them decide on their agendas. This is how they do it in almost all Middle Eastern parliaments. The commonwealth of learning has no one to impose an agenda on its members. Nevertheless, science progresses along given lines that win great consensus about its agenda. (With no elections and no obligations, the consensus has to be clear and so it has to be wide.) Most philosophers stress that science has consensus as to which theory is true. Perhaps. The consensus regarding answers is much less significant than the consensus regards questions. Though obvious, among philosophers only Russell, Popper and Bunge have asserted this—presumably, following the footsteps of Einstein.

Scientific questions are not new. Ancient books of problems (beginning with Aristotle or pseudo-Aristotle's *Problemata*) testify to that. The scientific revolution rendered those questions obsolete, and Robert Boyle replaced them with the extremely popular and now forgotten questionnaire *General Heads for the Natural History of a Country Great or Small Drawn Out for the Use of Travellers and Navigators*. One of its famous descendants is the famous *Notes and Queries of the Royal Anthropological Institute*. (See my *The Very Idea of Modern Science: Francis Bacon and Robert Boyle*, 2013, 181.) Nevertheless, the very idea of a question raises great ambiguity. William Robertson Davies, who wrote a few successful novels, reports in his most famous one (*The Rebel Angels*, 1981) that is centered on Academe, that experimenters say: I do not ask questions but simply record what is in front of my eyes.

The source of this expression is Bacon's inductive philosophy. He said, most astutely, questions are tools for selection. A selection rests on an idea. The idea is scientific and thus incontestable, or else it is a prejudice and thus an impediment to research. Hence, in the beginning science must be unselective. His posthumous book *A Forest of Forests* (*Sylva Sylvarum; or, A natural history, in ten centuries. Whereunto is newly added the History natural and experimental of life and death, or of the prolongation of life*, 1670) contains a lot of rubbish. It deserves praise nonetheless, as it as the first database ever. Already Boyle wanted to replace it. He called the replacement *The Promiscuous Experiment*. He worked on it for decades; he never published it. (A fragment of it appeared posthumously as *Experimenta et Observationes Physicae*.) What impeded him was a serious difficulty. He held that science begins not with pure data, but with generalizations, and he knew that these are always hypothetical. He did not want any hypothesis to have the status of certitude that Bacon had ascribed to pure data. The idea of a database arose again in James Spedding's book on Bacon (*Evenings with a Reviewer; or, a Free and Particular Examination of Mr. Macaulay's Article on Lord Bacon, in a Series of Dialogues*, 1848). He said, we should try out this idea. In a review of this book, the great William Whewell said, the Royal Navy has a huge database on the weather and it is worthless, as any database not theoretically informed must be. Today, when we have databases galore, it is evident that they must be theoretically informed, since these days the problem of their classifications and of the translation from one system to another engage many researchers. I will skip this as it is task-specific. The demand for a database for non-specific scientific use is present in the writings of Charles Saunders Peirce, who said, the assessment of the probability of a hypothesis must make use of all extant data or else it invites prejudicial selections. Even were this possible, and even were the theory of the probability of a hypothesis satisfactory, the question raised here would remain: how does the commonwealth of learning fix its agenda? Does it? Even if it does have an agenda, it must always be under discussion and review. The commonwealth of learning always has multiple and partially overlapping agendas; they come with varying degrees of recognized authority. As Polanyi has observed, authorities fix the research agenda; indeed, as Thomas Kuhn has observed, managing to fix an agenda is the acquisition of authority. There is no study of this complex and fascinating and very important phenomenon.

It is becoming to contrast the inductivist view of science with the instrumentalist view, as these are the two competing traditional views. Instrumentalists never discussed agendas. Even their own agenda they took for granted. They had a good reason for this: they were conventionalists and they took it for granted that decisions concerning agendas are purely conventional: people determine them rationally in accord with their own concerns and interests. Any critique of conventionalism should rest then on the matter of conventions in general, not of agenda in particular. The most obvious question on this concern an obvious fact: shared tastes are essential for the agreement to set conventions for cooperation. The question then is, why do so many people agree on so many items that they cooperate on? In particular, can there be a scientific etiquette for fixing agendas? If yes, what is its rationale? Conventionalists do not care about this question, as they leave it to the individuals concerned. Some students of the diverse social sciences, however, do care about this question.

Polanyi agreed that fixing of agendas might go beyond the concern of specific individuals or groups, as it may concern the commonwealth of learning as a whole. It is then an interesting and significant problem—even if it concerns only specific disciplines as wholes. Leading scientists in any given field to determine the agenda there. Most fields do possess agreed agendas. (Thomas S. Kuhn won great popularity when he declared the leaders responsible for changes in the research agenda of their fields that he renamed paradigm-shifts and scientific revolutions; he also identified "fields of expertise" with university departments that have recognized agendas; *The Structure of Scientific Revolutions*, 1962, Introductory Essay:

Structure.) How research agendas crystallize is an interesting problem. Einstein suggested that fixing a good research agenda requires a good metaphysical idea. (As these are rare, there are only a few options for serious researchers to follow.) Following this idea, I have suggested that the general agreed view of things helps decide such matters, for, fields that house different schools of thought tend to have different agendas, one for each school ("The Nature of Scientific Problems and their Roots in Metaphysics", in Mario Bunge, ed., *The Critical Approach: Essays in Honor of Karl Popper*, 1964).

Popper made a great contribution to the development of a theory of the agenda. Until he came along, philosophers had little to say on such matters, since they used to explain unanimity as the endorsement of successful theories. They thus scarcely discussed problems, much less the order of their significance—even while acknowledging that in some fields some problems have gained priority. The question of setting agendas is both theoretical and practical, and I will not discuss them here but rather the importance of the philosophical background to the political variant of the problem.

The only traditional item of setting the scientific agenda was the dispute between advocates of the views of science as a priori valid or a posteriori valid: the one recommended stress on theory, the other on experiment. As (Democritus and) Popper has pitched theory and experiment against each other, he pointed out that stress upon one or the other varies with circumstances.

The control of the agenda constitutes political power. This is obvious: every time a parliament opens, a real battle over the control over the steering committee takes place. In the commonwealth of learning this happens in international and national conventions. As the struggle for power there takes place behind the scenes, it is easier to discuss politics proper first. And to discuss the philosophical background to the political background of agenda making, it is useful to take recourse to Popper's political philosophy, as it is minimalist. We have here all sorts of parties, from the radical right that is quasi-anarchist to the radical left that is technocratic; in democracies, they share minimal democracy, and Popper's presents it as the ability to change a regime with no bloodshed.

Democracy is reformist and so it conflicts with both extremes: it opposes both radicalism and technocracy. It takes situations as given in attempt to improve upon them through critical discussions and votes. This recipe has its parallel in the commonwealth of learning. There traditionalism takes traditions as the fund of knowledge. Radicalism ignores traditions and considers knowledge in the abstract. It takes the knowledge in question to be the possession of "the" individual, and then it has the choice between the fiction of the Enlightenment that everyone is expert or the Platonic ideal of the perfect technocracy of the philosopher king. Critical rationalism takes knowledge as given and suggests improving upon it with the aid of critical discussions.

Critical rationalism is essential for democracy, and even for much less. For example, when the Soviet Union made efforts to soften its dictatorship in the brief period known as *glasnost* (openness) and *perestroika* (restructuring), it was dabbling with critical rationalism. The theory that only science is rationally justified raises the question, how is critical rationalism applicable to politics and can it allow for pluralism? How does the philosophy of science enter political philosophy? How does political philosophy enter the philosophy of science? In critical rationalism, this is hardly a problem: we can have it any way we want as long as we invite critical debates and are ready to try to improve. Traditionalism and radicalism are much less flexible.

Traditional philosophy considered the problem, the choice, the concern, all as individual: society had no role to play in them except to allow for them to this or that degree. Perhaps even the individualism that is so characteristic of western society, culture, philosophy, is but a by-product of the appeal of the Greek ethos to the individual—simply because it was an invitation for all to be critical of conventional wisdom. (We have a similar aspect in Hebrew tradition: the prophets attacked (not conventional knowledge but) conventional mores. They tended to appeal to individual conscience, having nothing stronger to fall back on.) Since problems and discussions center upon the individual, the social component of individual choice entered as a mere after-thought. This is not sufficient.

All this explains how, oddly enough, no traditional theory of rationality supports democracy, and that critical rationalism appears as its foundation very late in the day. Democracy cannot easily enter philosophy a posteriori except if it is an existing option, and appealing and strong at that. Where does society enter the discussion "naturally"? At the very start: at the very setting of its central problem. The fire burns the same way everywhere, said Protagoras, but the law is different in different societies. What of it? Why did this trouble the ancient Greek sophists so much? Because this is their discovery of almost limitless new options: the choice between competing traditions is open! Herodotus narrates that people overlooked these options because they "naturally" considered inferior all customs other than their own. That there was a blind spot that was to clear before this was perceivable, as he explained by narrating a myth: the magi had to be expelled before rationality could be instituted. Now all this may have been revolutionary in Antiquity and even in some not so distant past, but for most modern people with even minimal education it is today neither very new nor very exciting. However difficult it is for an individual to break ties with tradition, it is possible, and it is done, and at known cost, and while raising certain social and political and educational problems—not philosophical ones. Philosophically, every extant attitude to custom is rather unproblematic, because we have learned to circumvent them: out of so much frustration, perhaps, we have found that it is easier to attack a social problem by new technology than by forging and/or implementing new attitudes. This finding enters every agenda stealthily.

Attacking social problems by applying natural sciences to improve technologies rather than by applying social sciences to effect social reforms may also bring about postponement of treatment of the ill effects instead of appropriate solutions. The cost of this postponement rises with high interest rates. The ecological and the peace movement are discussing in such a great and colorful detail the coming cataclysm. All traditional societies, primitive, advanced, and in between, are (almost by definition) unable to deal with cataclysms of any sort. If the future is anything like the one that the ecologists predict, then we are doomed, and they only contribute to it by raising panic. We need urgently a discussion of the agenda of the ecological movement; they will not do it, and this raises an urgent question on the public political agenda: who should do it? More precisely, how should we interest politicians in it? I have no idea.

If there is any rhyme or reason in current global politics, it is its being so very marginal: almost all global political activity now serves local interests: politicians pay heed to global politics only through the prism of international politics or, worse, through national politics and even party interests. Few politicians ever have the resourcefulness and vision to develop long-range plans, anyway. Progress never-theless has taken place: politicians from time to time did recognize the requirement to come up with long-range plans and engage increasingly wider outlooks. Otherwise, it would not be possible to develop the modern nation-states out of small local interests. Once a politician saw the advantage of a wider outlook, of planning for larger systems, their ensuing success forced competitors to follow suit. This is no longer the case. If the whole world will not mobilize in order to put global issues high on their agendas, then nothing will ensue: only global solutions may possibly meet the global problems of the survival of humanity; local solutions will not begin to touch them. How then can the world leadership in all of its variety agree that certain moves are essential for global survival? Is there hope to alter the habits of millennia and invite more rationality in global politics before it is too late? Can there be a strict test for the environmentalist claim that the wide use of fossil fuel risks our very survival?

The present problematic situation is a case of an inability to apply to it known solutions. What is the situation from a critical rationalist viewpoint? There is a clear sense of crisis in the air. Whether the future of humanity is gloomy or not, ecologist tell us that it is under a severe threat. If survival is the topmost item on every practical agenda, then the survival of humanity should be the topmost item of every agenda, private, philosophical, national, global. Does any philosophy address this matter? Not that I know of. Philosopher Arne Naess (1912–2009) decided early in the rise of the ecological movement to devote all his energy to ecological problems. He failed to put ecology high on the global political agenda.

This is a matter for fixing both the social environment in which fixing an agenda is possible and for fixing a steering committee for that agenda. This essay began with first person singular. I do not know even what book to read next. The natural move is to my social milieu. How do I integrate within it? How does it determine and improve public agendas? In particular, how does the commonwealth of learning do that? I do not know and I invite any interested researcher to study this situation.

Chapter 3
Intentional, Unintentional and Sub-intentional Aspects of Social Mechanisms and Rationality: The Example of Commitments in Political Life

Alban Bouvier

Abstract In this chapter, I criticize recent programs and sub-programs in analytical sociology regarding the lack of attention they pay to the issues of rationality and intentionality. I put forward the idea that the rationalist paradigm in social sciences is not reducible to Rational Choice Theory. I also argue that establishing a dichotomy between intentional and unintentional processes is too simplistic: we also need to consider sub-intentional processes. Indeed, I focus in particular on interpersonal commitment (e.g. shipmates committed with each other to row as much efficiently as possible), a kind of social process that may be either intentional or sub-intentional. I then explore the empirical relevance of this conception of commitment by analysing several historical examples—borrowed from contemporary processes of decolonization and secession—of political commitments or alleged commitments.

Introduction

I would like to tackle, from within the framework of analytical sociology, two closely related issues in the social sciences: rationality and intentionality. I understand analytical sociology as aiming to *decompose* social phenomena into their elementary components (beliefs, preferences, emotions, relations of trust, relations of authority, social norms, systems of relations, organizations, institutions, states, etc.) and their elementary mechanisms (aggregation, coordination, cooperation, conflict, segregation, secession, etc.). This project includes the analysis of the

A. Bouvier (✉)
Aix-Marseille University, Marseille, France
e-mail: bouvier.alban@hotmail.fr

A. Bouvier
Institut Jean Nicod (CNRS/ENS/EHESS), Paris, France

© Springer International Publishing AG, part of Springer Nature 2018
G. Bronner and F. Di Iorio (eds.), *The Mystery of Rationality*,
https://doi.org/10.1007/978-3-319-94028-1_3

various sequences that may compose a mechanism (triggering events, domino effects, cascade effects, threshold effects, feedback effects, etc.). The idea of analytical sociology is closely connected with Robert Merton's ideal of constructing, step-by-step, "middle-range theories" or micro-models of phenomena, instead of, on the one hand, creating a global theory aiming at understanding all social phenomena under a set of very few basic concepts or, on the other hand, writing mere narratives (Merton 1967; Bouvier 2008).

Although the issues of rationality and intentionality have been investigated at length and in various ways by Thomas Schelling (1978a, b), Jon Elster (1979, 1983), Raymond Boudon (1982, 1994) and James Coleman (1990), who are often viewed as precursors of analytical sociology, these two related issues have been addressed, on the contrary, only cursorily in the *Handbook of Analytical Sociology* (Hedström and Bearman 2009b; Bouvier 2011b). Thus, in their introductory chapter, Hedström and Bearman (2009b) criticize the Hempel model of explanation in such an expeditious way that they entirely ignore Hempel and Dray's debate (in particular Hempel's response to Dray's objections in 1962). This debate concerned the role of "good reasons" and "generative mechanisms" in relation to Hempel's other models of explanation in social science. In a more recent publication, Hedström and Ylikoski (2014) establish a close connection between these issues of rationality and intentionality by harshly criticizing Rational Choice Theory. They argue that RCT assumes that actions are necessarily purposive (or "intentional") and contrast it with functionalist theories in psychology, in particular Festinger's theory of cognitive dissonance reduction (Festinger 1957).[1] In fact, the latter, according to the authors, turn out to be at least as relevant to the social sciences as the RCT. Within this context, they understand RCT so broadly that they include Boudon's theory, without mentioning that Boudon's (1996, 2003a, b) theory of "good reasons" goes far beyond RCT.[2] Indeed, the intentionality issue itself may not be quite as simple as Hedström and Ylikoski seem to think. There may be intermediary mental states between intentional and completely unintentional mental states; and we should perhaps make room for *sub*-intentional states and for *sub*-conscious intentions. Friedrich Hayek's work (1952, 1962) is potentially very fruitful here; although neglected in the recent social sciences, it has proved to be seminal in the cognitive sciences. Furthermore, there have been various interpretations of cognitive dissonance reduction processes and one of them is that the process at issue might be interpreted, at least sometimes, in terms of *commitments* (Heider 1958; Kiesler 1971) —that is as *intentional* or, perhaps more accurately, *sub-intentional* processes— instead of purely *unintentional* functional processes. These sub-intentionalist theories of commitments fit in well with Hayek's views.

---

[1] The authors also mention Tversky et al. (1980), Milgram (1963). One could probably add Sperber (1996, 2011).

[2] The standard version of RCT in sociology (e.g. Coleman 1990) stipulates that rationality is adaptation of means to ends (or pragmatic—or strategic—rationality) and that ends are material well-being (the "maximization of utilities").

Gianluco Manzo, editor of the volume in which the Hedström and Ylikoski paper was published, outlines a more open view of the scope of analytical sociology (Manzo 2014b) although there is no specific mention of the debate between Hempel and Dray nor of Hayek's idea of sub-intentional procedures. In particular, Manzo suggests not only that explanation by "good reasons" has a place within analytical sociology, but also that it is compatible with cognitive functional explanations (pp. 21–27). However, Manzo highlights a very specific sub-program of analytical sociology (p. 6), which explicitly emphasizes the role of agent-based computational modeling (pp. 7–10) and, generally speaking, formal modeling (in particular graph theory). While I do not contest the legitimacy of such a sub-program, I suggest that its specific focus does not address the issues of rationality and intentionality in detail. Furthermore, in mentioning that "the two theoretical pillars of analytical sociology" are "actions and networks" and that "analytical sociology is all about the complex interplay between 'actions' and 'networks' (and social structures more generally)" (Manzo, p. 6), Manzo emphasizes the place of network analysis to such an extent that social structures (which include organizations, institutions, States, etc.) seem reducible to mere networks. In other words, macro-sociology becomes reducible to meso-ciology or even to a certain kind of meso-ciology (meso-sociology not being entirely about networks). In this chapter, I will focus on the micro-level—what Manzo prefers to call "actions"—and, still more specifically, on how individual actions may become collective. I will also show the effect of individual actions and interactions at the macro-level of states, exploring how micro-events may trigger macro-events such as secession of states and decolonization processes, thereby examining the link between micro-sociology and macro-sociology (See Schelling 1978a; Bouvier 2011a). Meso-sociology, since it focuses on informal relationships and informal groups as well as intermediate institutions, will fit in this three-tiered framework.

Independently of the specific debate on the issues of intentionality and rationality I have discussed, the concept of commitment is interesting more generally speaking. Recently, it has been one of the most widely used in the social sciences, although with various meanings. We see it in the social psychology of Heider and Kiesler, the economics of Amartya Sen (1977) and the social philosophy of Margaret Gilbert (1989), to take only a few examples.[3] However, despite this interest, the phenomena that the concept of commitment is meant to describe have not yet been investigated from the perspective of analytical sociology (with the partial exception of Schelling and Elster's examination of pre-commitments). I maintain that the concept of commitment deserves at least as much attention as that of trust, which has been the object of much work in analytical sociology (Gambetta 1988, 2009; Coleman 1990; Cook and Gerbasi 2009).

The widespread interest in the concept of commitment—and in particular interpersonal commitment—in the social sciences is understandable by the fact that

---

[3]For a useful survey, see also Peters and Spiekermann (2011).

commitments may be a better "cement" of sociality,[4] along with the more passive interiorization of norms via processes such as *habitus*,[5] than the mere calculation of material self-interest. Symmetrically, the *lack of commitments* may greatly impede the functioning of social and even economic life.[6] Furthermore, *violations of commitments* may be one of the sources of major violent conflicts. In fact, violations of commitments in particular may act as *triggering events* to a chain of almost unavoidable other events—that is, as components of particularly obvious social *mechanisms*.

The first section will be devoted to set forth the relevant general philosophical background necessary to ground the issues of rationality and intentionality within an analytical sociological framework. In the second section, I will investigate intentional commitments in relation to the issue of rationality. In the third section, I will describe interpersonal commitments that involve sub-intentions, and thereby challenge those authors who set up a simple contrast between intentionalist and purely unintentionalist theories. In the fourth section, I will outline an analysis of historical case studies illustrating the processes theoretically investigated in the third section.

Generative Mechanisms, RCT and Good Reasons. *Conscious, Unconscious and Sub-conscious Rationality*

In this first section, I plan to clarify two fruitful ideas: the idea of explanation by "good reasons" and the idea of explanation by referring to sub-conscious or unconscious processes.

Raymond Boudon is commonly viewed as one of the precursors of analytical sociology for two main reasons: he devoted specific attention to Merton's idea of middle-range theories (Boudon 1991); and he tried to investigate certain elementary generative mechanisms of social inequality in modern societies (Boudon 1974, 1982). Boudon was also a strong critic of overly narrow conceptions of rationality in the social sciences, illustrated for example by Coleman's model of rational action (Coleman 1990). Boudon suggested, without fully exploring in a philosophical manner, the relevance of two ideas: the idea of "good reasons" (1996, 2003a, b)—at first glance trivial—and the idea of "meta-conscious" processes (Boudon 1994, 1995)—at first glance obscure. I would like to show, first, that it makes sense to distinguish between three levels of mental states: conscious, sub-conscious, unconscious and, as a consequence, between intentionality, sub-intentionality

---

[4]In *The Cement of Society*, Elster (1989) did not investigate the role of interpersonal commitments as elementary components of social life.

[5]I have no room to discuss Bourdieu's theory in this paper but one could easily argue that Bourdieu always hesitated on the unintentional or the sub-intentional nature of *habitus* (Bourdieu 1990).

[6]See Sen (1977).

(as subconscious intentionality) and absence of intentionality. Second, I would like to argue that it makes sense to speak of rationality in terms of good reasons, that is, of rationality in a much broader and more general sense than in the standard, and overly restrictive, versions of RCT that rely on the aforementioned two levels. For pedagogical reasons, I will investigate these two issues in reverse succession, with the second point before returning to the first.

(1) RCT, good reasons and mechanisms

I will begin by revisiting a nowadays forgotten debate between Hempel and Dray on the kinds of explanation relevant in the social sciences. Boudon has used the notion of "good reason" a great deal in his explanation of many social phenomena since at least *The Art of Self-Persuasion* (1994) to *Raison, bonnes raisons* (2003). However, he only acknowledged the work of Dray in a very brief footnote and without referring to the latter's debate with Hempel (Boudon 1994, p. 203, n. 38).[7]

In 1957, Dray (1957) had contended that explanation by actors' good reasons (or "understanding") was the relevant method of explanation in the social sciences and especially in history. Contrary to what is often assumed, Carl Hempel (1962) largely agreed with Dray on the role of explanations by good reasons in social science. Where they disagreed concerned the compatibility of this kind of explanation with other modes of explanation. Indeed, for Hempel, explanations by good reasons could function in principle *within* the inductive-statistic model (largely used not only in the natural sciences, but also in demography, sociology and economics) and even within the nomologico-deductive model.[8] Hempel argued that convincing explanations by good reasons were based on the observation of regularities in similar contexts—which is precisely what the inductive-statistic model requires—although these observations are generally far too rare to permit rigorous statistical correlations.[9]

Hempel (1962) also referred to the "logic of the situation" (p. 28) in the sense of Popper's "situational analysis" (Popper 1957) to specify what good reasons are: reasons that seem empirically and logically valid to a social actor *within a certain situation*—a situation that is always local and partial,. In fact, most of Boudon's analyses, since at least *Le juste et le Vrai* (Boudon 1995) can be read as brilliant empirical applications of Popper's situational analysis to a number of empirical case studies in sociology, in particular in the sociology of moral norms.

Let us define explanation by good reasons more precisely. Hempel quoted Dray: this is the "reconstruction of the agent's *calculation* of means to be adopted toward this chosen end in the light of the circumstances in which he found himself" (emphasis is

---

[7]The probable background of Boudon's intuitions in Dray has been mentioned by several authors [e.g. Nadeau (1993) and Di Nuoscio (1996)].

[8]According to Hempel, the inductive-statistic model of explanation is itself formally reducible to the nomologico-deductive model of explanation. This is why one can speak of a "unique" model.

[9]Hedström and Swedberg (1998) mentioned the 1962 Hempel paper (p. 8), but retained only the ideas dating back to Hempel (1942) and displaying the nomologico-deductive model.

Dray's) (p. 25).[10] Certainly, most historians would agree with this. However, Hempel added that explanation in history and the social sciences requires—more generally— the search for generative mechanisms (Hempel spoke of "genetic explanations in history", pp. 21–25).[11] In these few paragraphs, Hempel outlined almost the entire program of contemporary analytical sociology.[12]

To conclude, I am ready to acknowledge that the issue of knowing whether one should uphold a much larger view of RCT than the standard version (by broadening the meaning of "rational" in "Rational Choice Theory") (Opp 2013) or go "beyond" RCT (Boudon 2003a, b) is—to some extent—a lexical issue, as Opp (2013) himself suggests. However, because the concept of "rational" in "RCT" is generally defined rather narrowly, it is probably preferable to forego "RCT" as a label altogether. Since "explanation by good reasons" may seem slightly too trivial, I propose instead the term "rationalist paradigm." This wording is used in history of science in this sense[13] and refers to what could be called "epistemic rationality," or "cognitive rationality" as Boudon put it, on suggesting that means-ends rationality requires information (or knowledge) about the ends, the means available, the adaptation of these means to the ends aimed at, etc.[14]

(2) Meta-conscious or sub-conscious processes *versus* conscious and unconscious processes

Let us go to the issue of intentionality.

Von Mises (1949) was the main supporter of a radical demarcation line between "actions" (by definition intentional) and "behaviors" (by definition unintentional), and of the correlated idea that social sciences only deal with "actions" (while "behaviors" deal with natural sciences). However, Hayek (1952, 1962), unlike von Mises (who taught Hayek), introduced the idea that one should leave room to what

---

[10]Hempel also quoted Gardiner (1952, p. 136): "In general, it appears safe to say that by a man's 'real reasons' we mean those reasons he would be prepared to give under circumstances where his confession would not entail adverse consequences to himself" (Hempel 1962, p. 30). This is what many other authors (e.g., Boudon 1989, 1996, 2003a, b), call: the *reconstruction* of plausible reasons.

[11]Hempel (1962) takes examples of "genetic explanations" in physics; the simplest one is of a falling stone (p. 23).

[12]Hedström et al. (1998), rightly emphasized the proximity between Popper and Rational Choice Theory. To my knowledge, Hempel has not been read yet with as much attention as Popper from the viewpoint of analytical sociology. I fully agree on this point with Opp (2013). Demeulenaere (2011) rightly noticed that sometimes in Hempel's accounts covering-law explanations "include mechanism-based explanations" (p. 190) but did not mention the 1962 paper and did not comment either on the Hempel-Dray debate.

[13]See, e.g., Koyré (1971).

[14]Among the ends, we also need to acknowledge the preferences that social actors do not see as reducible to their self-interest or their personal preferences, but instead as independent objective values. Boudon prefers to refer to Weber's notion of *Wertrationalität* (axiological rationality) as *opposed* to *Zweckrationalität* (means-end rationality), but the general idea is the same. See section II.

he called "meta-conscious" or "supra-conscious" processes, namely rules guiding perception, knowledge and actions.[15]

One interpretation of Hayek's work is that these meta-conscious processes are radically unconscious because they deal with neurological processes. Another interpretation, especially based on Hayek (1962), is that there is continuity between conscious processes and certain unconscious processes and that the concept of meta-conscious states aims at accounting for these intermediate states.[16] I will retain this second interpretation (the continuity interpretation), which essentially states that rules are not radically unconscious, while assuming that there are also radical unconscious processes, namely the neurological processes. But if (a) one generalizes the idea of continuity between consciousness and other mental states beyond the case of rules and (b) one focuses on the idea of *degrees* of consciousness, it seems more appropriate to speak of *infra*-consciousness or *sub*-consciousness when one thinks of a low degree of consciousness rather than of meta-conscious or supra-conscious processes.[17]

Boudon explicitly borrowed the wording "meta-conscious" from Hayek, with whom he also shared a similar conception of general rules or "a priori."[18] However, implicit in Boudon's conception of the meta-conscious are not only general rules but also more particular and contingent sub-conscious processes. In fact, according to Boudon, what an actor clearly identifies *after the fact* as the reasons or intentions behind a particular action, often dismissed as post-factum justifications, could have nevertheless sub-consciously motivated the action in question. This gives rise to the idea of the progressive emergence of intentions or of "proto-intentions" (see third section).[19]

To conclude on this point, the search for generative mechanisms can include not only purposive rational choice (like in Coleman's work), nor even more generally intentional good reasons, but also *sub-intentional good reasons*. Of course, this does not exclude the rationality of *unintentional processes* aiming at the survival or the welfare of living being—a kind of rationality that is very limited and even "myopic" (Elster 1979, 2007). Thus, a purely unintentional mechanical processes of reducing mental dissonances to keep one's mind quiet might be functional (although only weakly). But here I have wanted to highlight alternative theories.

---

[15]See in particular, Hayek (1952), Chap. 6.6 and Hayek (1962). See also Di Iorio (2015) on this dimension of Hayek's work and on the intellectual relationship between Hayek and von Mises.

[16]On these two interpretations, see Fleetwood (1995), Chap. 8. On the second interpretation, see J. Gray (1984), Chap. 2. See also Di Iorio (2015), Chap. 2.8.

[17]The idea of supra-consciousness might suggest a higher degree of consciousness. I mean the exact opposite.

[18]See Boudon (1994[1990]), p. 110, Boudon (1995), p. 138–139, n. 4.

[19]Boudon is not explicit on this point. On the more general relationship of Boudon to Hayek, see again Di Iorio (2015) especially Chap. 5. What Manzo (2014b) wrote on recent cognitive theories, in particular the program he has formulated, is also in line with Hayek's views on "meta-conscious processes" (pp. 25–26). However, Manzo tackled neither the issue of the distinction between sub-intentional and un-intentional processes, nor the issue of the emergence of intentions from proto-intentions.

Intentional Commitments, Rational Choice Theory and "Good Reasons"

In this second section, I compare the extent to which the broader conception of explanation by "good reasons" can account for commitments with RCT in its standard version, that is, a very narrow version of rationalism. I first discuss Elster's account of pre-commitments and then set forth two distinct but correlated accounts (Gilbert's and Sen's) of what I take as proper commitment: interpersonal commitment including the idea of obligation or duty. In both cases, I show the relevance of the idea of explanation by "good reasons".

(A) *Pre-commitments*

Elster (1979, 1983), following Thomas Schelling (1978b), described commitment as "deliberate shaping of the feasible set for the purpose of excluding certain possible choices" (1983, p. 114). Homer's telling of Ulysses and the sirens in *The Odyssey* is a paradigmatic example of this mechanism. Ulysses is afraid of being ensnared by the sirens' song as he and his shipmates come near sirens' realm and he is afraid of having their ship destroyed by the reefs on which the sirens rest. Nevertheless, he would like to listen to their famed and entrancing song. His strategy is to ask his shipmates to bind him to the ship's mast and to request them to fill up their own ears with wax. This way, not only will they not be able to hear the sirens, but they will also not hear Ulysses' own quite predictable demands, upon hearing the sirens, to be untied and to direct the ship ever closer to the reef.

Ulysses *intentionally* limits his further choices. According to Elster, on the one hand, Ulysses is rational in the sense of means-end or "strategical" rationality, since Ulysses uses means (self-binding) adapted to his aim (listening the siren's songs); but, on the other hand, Ulysses is irrational since self-binding limits his opportunities by increasing his constraints. Ulysses is therefore partially or "imperfectly" irrational (Elster 1979).[20] One could nevertheless add that what may seem irrational here may be irrational only in the short term and may turn out to be rational in the long term: hearing sirens' songs *expands* opportunities—though aesthetic, rather than material opportunities—and therefore this kind of opportunity already requires a broader version of RCT, including aesthetic well-being.[21]

Elster (1979) not only gave examples of *individual* self-binding domains such as addiction to tobacco, gambling, etc., but also examples of *collective* self-bindings. Thus, the existence of political constitutions may be explained as a way of *pre-*

---

[20]Elster (1983) makes another distinction, between a "thin theory" and a "broad theory". According to Rawls, quoted by Elster, "the thin theory of the good [explains] the rational preference for primary goods" (this fits in with the standard narrow version of RCT) while Rawls acknowledges "that a fuller theory is needed to account for 'the moral worth of persons'" (Elster 1983, p. 1, n. 1).

[21]Thus, a even fuller theory would be needed: one that would not only take into account moral values (see previous footnote) but also aesthetical values.

commitments for Members of Parliament: when MPs will discuss about new laws, there will be bound by the fundamental constitutional laws they cannot change without resorting to very specific assemblies (e.g., in France, the Congress, which is composed of both Members of Parliament and Senators, has to be convened). And the reason why there is a distinction between fundamental laws (written in the Constitution), whose change requires a long process, and other laws is to prevent MP's of changes that would be motivated by too much contextual (and myopic) intentions.

To conclude, one should notice that pre-commitments, described by Elster as self-bindings, are not proper "commitments", which involve—in the usual meaning of the term—the idea of obligation or duty. Ulysses is not under any *obligation* towards anybody. But his shipmates, on the contrary, have *obligations* towards him and are *committed* with him, what Elster did not specify. This is what we have to investigate a little more.

(B) *Unilateral, reciprocal and joint interpersonal commitments*

Let us come back to Ulysses' fable. Although Homer does not seem to suggest anything on this point, Ulysses' shipmates were probably not only *unilaterally* committed with Ulysses to fulfill Ulysses' ends but they were probably also *reciprocally* (or even—see below—"jointly") committed 'with' each other to row more efficiently. I will comment in succession on these two notions: reciprocal commitment and joint commitment.

As seen previously, Elster interpreted certain behaviors in a Parliament as self-bindings and not as interpersonal commitments. But there are also interpersonal —and even reciprocal—commitments in a Parliament. In a Parliament, the majority does not only bind the minority and the minority is not only bound by the majority (as Elster emphasizes). MPs are also *reciprocally committed* "with" each other—to abide by the laws that they have voted for and that will deal with everyone's daily life (e.g. civil code, penal code and taxations).

Moreover, when there is a "we-intention" (Tuomela 2002), that is, an intention of playing one's part in the group (e.g. in a soccer team or in a political party involved in an electoral campaign) and of desiring the success of the group more than one's own personal success, Margaret Gilbert (1989) suggests that we speak of "*joint commitments.*" Joint commitments are significantly different from reciprocal commitments. Whereas reciprocal commitments are conditional, joint commitments are not: even if one person does not play his/her role, everyone else is still supposed to fulfill their functions and, in addition, compensate for the defective actor. This means that the "good reasons" to act that way are not material *self*-interest but instead the *group*-interest. This does not exclude the possibility that one obtains personal satisfaction by having acted in that way; however, this satisfaction might be only psychological—the satisfaction of having acted well—and not material. This goes again beyond a narrow version of rationalism such as the standard version of RCT.

In political life, reciprocal—conditional—commitments are frequent, while joint commitments are probably rarer. In a political campaign, the team's members are expected to work for the success of the leader and, as a consequence, to try to play their parts as team members, sometimes without considering rewards (excepting the satisfaction of having acted well). Reality is often less idyllic, since players generally expect rewards and are often tempted to play much more individualistically (like in every group, free-riding is frequent).

Besides, joint commitments in Gilbert's sense, although not conditional, are not necessarily moral either. To take an extreme example, certain Nazis might have felt jointly committed with each other to the extermination of Jews. Thus, the "good reasons," epistemologically speaking, for people to act in a certain way, even when they aim at the collective good for the group of which they are the members, may be "bad reasons" morally speaking. The notion of commitment, according to this account, requires a notion of duty that is formal, and not reducible to the more substantial notion of moral duty.

Amartya Sen introduced the concept of commitment completely independently from Gilbert and Elster, albeit within the same framework as Elster, that is a narrow (or "thin") version of RCT. It is well known that Sen's concept is not well defined, its content varying depending on the examples he examines.[22] Nevertheless, at least one set of these examples deals with Gilbert's understanding of joint commitments, in particular when Sen (1977) suggests that the British economic system did not function well in the 1970s because most people in the UK prioritized their own material self-interest without considering the public good. In other words, if people look for the public good, they may have the same "good reasons" for doing that that in Gilbert's contexts. In other contexts, Sen uses "commitment" and "(moral) duty" interchangeably, which demonstrates that he thinks of rationality in the sense of Rawls' and Elster's broad theory of rationality (see above). This is also the case of Boudon (1995), who speaks of "axiological reasons" (*Wertrationalität*) in analyses parallel to Sen's and closer to Kant than to Weber.[23] The main point here is that good reasons in all these contexts (aesthetical, substantially moral like in Sen's contexts, only formally moral like in Gilbert's contexts) are not reducible to RCT in its narrow standard version. However, in all these cases, commitments are supposed to be intentional and therefore fully conscious. We now need to go even further in the examination of the intentionality issue.

---

[22]See Pettit (2005) for example.

[23]See above, footnote 14.

Sub-intentional *"Commitments"* and Sub-conscious Good *Reasons in Interactive Contexts.* Theoretical and Experimental Perspectives

In this third section, I tackle more subtle and more complex kinds of commitments than those addressed previously. I want to show that, even in these cases, it still makes sense to think of rational actions in terms of good reasons. I will speak of sub-intentional *commitments* as good examples of phenomena requiring a "sub-intentionalist" explanation. The aim is to show in detail that alternatives to purely un-intentionalist explanations work well in certain situations. More precisely, I will try to show that certain theories in social psychology may allow us to go further than Margaret Gilbert on the phenomenological accounts of commitments.

Instead of focusing on explicit joint commitments like those that may occur in Parliaments (or in private juridical contracts), Margaret Gilbert has examined the *implicit or tacit joint commitments* that, she argues, often happen in the simple circumstances of everyday life. These implicit or tacit joint commitments are not expressed by any explicit wordings but are nevertheless common knowledge. They can occur when a member of a group decides to express a view as the group's view only because nobody disagrees explicitly ("silence means consent") (Gilbert 1989). It is easy to hypothesize that, if the supposed commitment was tacit, there might have been some misrepresentations regarding the exact content of the commitment and even on the reality of this alleged commitment. As a consequence, in these situations, people may feel entrapped. [Of course, these tacit commitments may also frequently occur in political life, and this is what I will highlight in the following section (IV)].

One can go further on the investigation of the nature and role of commitments by referring to social psychologists who have identified how people can feel committed without having clearly wanted to commit. Charles Kiesler's theory (Kiesler 1971) is based on the idea that people can feel committed post hoc to support the general principles that can explain a choice they have made earlier *without thinking thoroughly* about the implication of their initial choice. There is a thin line of demarcation between, on the one hand, believing that we are committed with full awareness and, on the other, feeling entrapped by someone else who may have wanted to lead us where we did not want to go or where we are not sure now whether we had earlier wanted to go. The distinction between these two mental states depends on whether one can recognize the reasons that could have rationally motivated our actions (of which one was not fully aware initially) as the real reasons of our actions.

Kiesler (1971), elaborating on experiments on the foot-in-the-door techniques used originally as a marketing "trick" (Freedman and Frazer 1966), has argued that *unilateral* and *sub-intentional commitments* (on the part of potential customers, for example) may be not only rationally—in the sense of the RCT (that is strategically) —induced (in the case of market: by sellers) but also rationally justified by good reasons (of customers) and assumed post hoc. This phenomenon requires further investigation.

Regarding the *foot-and-the-door technique*, Freedman and Frazer (1966), in very famous (but still under-exploited in the social sciences) experiments, asked people to place a small card in a window in their home supporting safe driving, without revealing that they were psychologists conducting an experiment. About two weeks later, the same people were asked by a second person to put a large sign advocating safe driving with the same message ("Drive Carefully") in their front yard. Experiments reveal that this is statistically much more efficient than to simply ask people to put the large sign in the garden from the outset. Functionalist accounts in line with Festinger's theory would say that this is the outcome of a merely mechanical *unintentional* process: human mind has an inner drive to make one's beliefs and actions coherent in order to avoid disharmony.

Kiesler's account is alternative. This is a phenomenological account (it describes intentional or sub-intentional processes). Kiesler argues that in these experiments, people might have felt *retrospectively* that they did not only commit to a single act when they initially accepted to place the small card in their window but in fact to a much more general campaign, even if this retrospective feeling was purely *subjective,* since such an *explicit* commitment was not required at all and they did not explicitly agree on anything of the sort. One can add to Kiesler's analysis that in both cases, if Kiesler is right, what was at issue as the bases of these behaviors was *not* material self-interest, since they did not receive any financial or material advantage: rather, people acted as they did because they think it was their duty (Sen). Regarding the second case, moreover, Kiesler argues convincingly that the people might have felt committed (or *jointly* committed, per Gilbert's theory) with the persons (actually psychologists) who seemed to act in favor of the safe driving campaign.

Furthermore, Kiesler adopts a rationalist point of view and assumes that people plausibly reconstruct their *sub-conscious reasoning* (good reasons) as follows. If I agreed to put a card in my window, it is in fact because I supported their road safety campaign (displaying a card being a logical consequence of agreeing with the principle of the safety campaign). However, since displaying a large sign advocating safe driving is also a logical consequence of the same general principle, if I have agreed to this principle, I should also agree to this new consequence and therefore I should agree to put this sign in my front yard. Of course, this reconstruction may be wrong and it may be only a "rationalization" that does not fit the genuine process. But, it is also plausible that the motivation in question was really there and simply not yet clearly recognized or assumed as intention, since it was still a sub-intention—or a "proto-intention" (see first section).[24] What could have been at work was an *implicit* sub-conscious or sub-intentional commitment (a "proto-commitment").

---

[24]Kiesler does not use this notion, which I find relevant here. See Romdenh-Romluc (2013).

In these cases, often people feel *entrapped* (a) because they realize that they are internally constrained to do what they did not initially want to do (or, at least, did *not quite consciously* want to do)—and (b) possibly because, in certain cases, they may have good reasons to suppose that someone else had, on the contrary, the quite *conscious intention* to induce such a process. The problem here is *not* that people did not anticipate the consequences of their initial act, but that they did not correctly grasp the true mental content of their *own* acceptance to do what they were asked to do, or the possible interactive context of their first choice. They may even not be quite sure that they had accepted anything else than what was explicitly asked (in this case, placing a small card on their window).

Sub-intentional "Commitments" and Sub-conscious Good Reasons in Interactive Contexts. Outline of Historical Case Studies: The Role of "Populism" in the Secession of States and the Process of Decolonization

In order to prove the relevance in sociology of these theoretical and experimental analyses, I would like to briefly sketch the analysis of a few historical political examples in which people might have felt entrapped by a certain equivocation on their commitments and where sub-intentions (or proto-intentions or proto-commitments) may have even played a significant role. In these examples, commitments or violations of commitments or alleged commitments—micro-sociological events—either have really acted or could have acted as *triggering* events of a chain of other events— leading to possibly tragic events at the macro-sociological level.

The most obvious cases of commitments-entrapments in political life are those in which people are entrapped by a leader or by activists. One can argue, for example, that many Nazis—or collaborators in occupied countries—did not realize entirely to what extent they were committing when they accepted to do something that was asked by a member of the Nazi Party and that seemed trivial at the time, such as the circulation of a tract. They may also not have been entirely clear about their own intentions (still at a sub-intentional and proto-intentional stage). Festinger's theory (Festinger 1957), Milgram's theory of submission to authority (Milgram 1963) and Asch's theory of submission to social conformity (Asch 1951) were initially formulated to try to explain how extreme despotism may have emerged in a country such as Germany (or France). The same theories could be used also to account for explaining the acceptance of Stalinism or Maoism. Alternative theories, focusing on sub-intentional processes (Heider 1958; Kiesler 1971, etc.) have emerged in the same general context.

But I would like to focus on the symmetrical phenomenon, which could be called "populism," when a leader (or several leaders) feel entrapped by a political body to go further than he/she initially wanted because—either intentionally or sub-intentionally—he/she had given signs that could be interpreted as commitments. In a sense, I am looking here for leaders who were not only self-bound, like Ulysses, but were (or were supposed to be) also jointly committed to their "mates", unlike Ulysses. I have examined the complex case of Stephen Douglas's accusations elsewhere (Bouvier 2016). Launched during the famous 1858 *Lincoln-Douglas Debates*, these accusations addressed Abraham Lincoln's supposed joint commitments with the Republican Party to immediate and radical abolitionism (while, in fact, Lincoln was a moderate on this issue and Douglas was aware of this) and can be linked to both the triggering of the Secession of pro-slavery States and the Civil War following Lincoln's election as President in 1860. Here, I will examine two or three simpler cases.

My first example is the decolonization of Central Africa in the 1960s. It has been argued that the process of decolonization occurred too rapidly to be fully efficient and that, more generally, many problems of the problems experienced by African countries in the 20th century can be traced back to rapid decolonization. This is not only a Western view that emerged post hoc to explain the failure of decolonization, but also one expressed by some of the main contemporary African political leaders during the process of decolonization itself. Although these leaders desired the independence of their countries, the most lucid among them doubted their respective country's ability to make a rapid transition. These political leaders knew that they needed a much higher political, economical, judicial education to build their nations—and perhaps most importantly, to avoid being cheated by Western political leaders during negotiations. However, as soon as they had put "the foot in the door" of the decolonization process, the African leaders were "entrapped" by the people of their own countries. More accurately: they had been entrapped by what could be seen by the people as their own previous commitments with the nation they belonged to. The case of Belgian Congo is well documented on this point. *Moïse Tshombe*, one of the leaders of the independence and a rival of the more charismatic Patrice Lumumba, upheld that he wanted independence but "pas aujourd'hui" ("not today") because he did not want, he said, to be pushed to take decisions "under the public pressure" (Van Reybrouck 2014[2010], Chap. 6). But this might have been not only a matter of *external* public pressure, but also—more subtlety—a matter of an *internal* mental process (by the sense of obligation inherent to commitments). Moreover, as (a) these "commitments" were not written and (b) leaders' intentions were not entirely clear for the leaders themselves—they might still have been at a sub-intentional and proto-intentional stage—certain leaders, like Tshombe, finally accepted willy-nilly to run the accelerated process despite (or perhaps due to) their feeling of entrapment.

Another dramatic episode of the same period, better known in many Western countries, is the case of *Charles de Gaulle's* alleged commitments during another decolonization process, this time in North Africa. Unlike the previous example, the sequence of events here involves only one leader, de Gaulle, who was the representative not of a colonized country but of the colonizing nation, France. In this case, we will consider in particular two very famous historical speeches that could have been understood by the audience as commitments. Indeed, when Charles de Gaulle did not follow through on these understood commitments, he was effectively accused of having violated them, what the Congolese leaders precisely wanted to avoid. First, on June 4, 1958, de Gaulle, then the President of the French Republic, stated to an Algerian-European audience in Algiers, an audience that did *not at all* desire the independence of Algeria, "Je vous ai compris" ("I have understood you"), as if to say he had grasped their (collective) will. Two days later, at Mostaganem (a city close to Oran), in another very famous speech, De Gaulle exclaimed: "Vive l'Algérie française!" ("Long live French Algeria!"). Most listeners thought he was committing (jointly committing) with the European population to keep Algeria as a French department, although what was occurring might have been a typical example of what happens in everyday life: there was equivocation whether it was or was not a commitment and, if so, on the exact matter of this commitment, since it was not a formal commitment at all (Horne 1977). De Gaulle might have been entrapped by the crowd's excitement and thereby felt compelled to use an ambiguous formulation. Nobody really knows—although many have claimed to—what De Gaulle wanted nor even if he was himself clear on his own intentions, which might have been at this time still only sub-intentional, alternating between several options and dissonant proto-intentions (Baumel and Delpla 2006). Whatever his intentions may have been, the feeling shared by numerous people that de Gaulle later violated this commitment because of a cynical *strategic* rationality and an entirely conscious intention *triggered* the violent reaction of the partisan of French Algeria, leading to a putsch attempt by general officers in 1961, in Algiers, to an assassination attempt against de Gaulle in 1962, at the Petit-Clamart, near Paris, to an amplification of the gap between Algerian-European and Algerian-African people and finally, as a counter-productive effect, not only to the independency of Algeria but also to the unavoidable exodus of almost all the white community ("pied-noirs") from Algeria to France.

There is no doubt, on the contrary, that when, on July 24, 1967, De Gaulle claimed in Montréal, "Vive le Québec libre! Vive le Canada français!" ("Long live free Quebec! Long live French Canada!"), this was deliberate, fully intentional and in accordance with the mythic vision that De Gaulle had of the greatness of France (Peyrefitte 1997). It could have triggered the secession of Québec, but it did not. We may trace De Gaulle's pronouncement to a myopic rationality regarding the international relationships (seriously cooled down between France, Canada and many other countries after this speech), although it admittedly reinforced the visibility of Quebec in the world. In any event, De Gaulle could not fulfill what could be interpreted as a commitment to support the secession of Quebec because he retired two years later. Finally, the Prime Minister of Québec, Daniel Johnson, despite being in favor of Québec's sovereignty, played the role of Tshombe in Congo: he did not want to be *entrapped* by the pro-independency Québécois, who had been galvanized by De Gaulle, and instead successfully (unlike Tshombe) slowed down the process of independence (Thompson 1990).

Conclusion: Analytical Sociology, Goods Reasons and Rhetorical History

In this paper, I have argued for the importance of commitments as components of social phenomena and for the relevance of the issues of rationality and intentionality, as understood respectively by Hempel and Hayek. In my opinion, these issues have been far too neglected in recent books devoted to analytical sociology. That is not to say that every behavior and belief is fully rational or is fully intentional but rather that issues of rationality (in a broader sense that in RCT) and intentionality (or sub-intentionality) are pivotal in the social sciences at the micro-level of explanation. We must leave room for them, so to speak, in the social sciences: many actions, in particular, may have sub-intentional sources but nevertheless be rational from a certain viewpoint. Nevertheless, I entirely acknowledge that this does not reject the possibility of entirely unintentional processes. Finally, I have tried to show—by outlining the analysis of several processes of decolonization or secession —how macro-sociological phenomena (concerning, for example, states or nations) such as civil war, secession, decolonization, may be triggered by micro-sociological events. This does not preclude the possibility that these micro-sociological events themselves involved not only individuals but also meso-sociological networks and entities such as informal groups (e.g. crowds in Congo, Algeria or Canada) or institutions (e.g. the emerging Republican Party in the US).

The absence of a careful investigation of these issues in the *Handbook of Analytical Sociology* may explain why the chapters devoted to history and to anthropology in this volume are so frustrating. The room left for history and ethnography is so meager that these "perspectives" are only presented in a fourth section as coming "from other fields and approaches," as if the sociological analysis of mechanisms itself should not include—as such—an historical dimension.[25] A sub-program at least, crossing Manzo's program for example, in the continuity of historical sociology, is both legitimate and needed (see, e.g., Elster 1990).

Moreover, in the continuity of my short case studies, I would like to specify that if one wants to investigate the entanglement of intentional and sub-intentional unintentional "good reasons" in the dynamic of social phenomena, emphasis should be given to what could be called "rhetorical history".[26,27]

---

[25]See, in particular, the two last chapters: "Analytic Ethnography" and "Historical Sociology", respectively by Diane Vaughan and Karen Barkey.

[26]On rhetorical history, cf. Zarefsky (1998) and Bouvier (2016). Zarefsky devoted many publications to the *Lincoln-Douglas Debates* (see, in particular Zarefsky 1990, 2014). On both the scope and limits of Boudon's theory itself on this matter, let me mention Bouvier (2007), and, on the more general issue of the relationships between Argumentation Theory and Rational Choice Theory, Bouvier (2002).

[27]I would like to thank Edward H. Barnet and Raphaël Künstler for their suggestions on the penultimate version of this paper.

References

Asch, S. E. (1951). Effects of group pressure upon the modification and distortion of judgments. In H. Guetzkow (Ed.), *Groups, leadership and men*. Pittsburgh, PA: Carnegie Press.

Baumel, J., & Delpla, F. (2006). *Un tragique malentendu, De Gaulle et l'Algérie*. Paris: Plon.

Boudon, R. (1974[1973]). *Education, opportunity and social inequality: Changing prospects in western society*. New York, Wiley.

Boudon, R. (1982[1977]). *The unintended consequences of social action*. London: Macmillan.

Boudon, R. (1989). Subjective rationality and the explanation of social behavior. *Rationality and Society, 1*(2), 171–196.

Boudon, R. (1991). What middle-range theories are. *Contemporary Sociology, 20*, 519–522.

Boudon, R. (1994[1990]). *The art of self-persuasion: The social explanation of false beliefs*. London: Polity.

Boudon, R. (1995). *Le juste et le vrai. Etudes sur l'objectivité des valeurs et de la connaissance*. Paris: Fayard.

Boudon, R. (1996). The cognitivist model: A generalized rational-choice model. *Rationality and Society, 8*(2), 123–150.

Boudon, R. (2003a). Beyond rational choice theory. *Annual Review of Sociology, 29*, 1–21.

Boudon, R. (2003b). *Raison, bonnes raisons*. Paris: P.U.F.

Bourdieu, P. (1990). *The logic of practice*. Cambridge: Polity.

Bouvier, A. (2002). An epistemological plea for methodological individualism and rational choice theory in cognitive rhetoric. *Philosophy of the Social Sciences, 32*(1), 51–70.

Bouvier, A. (2007). An argumentativist point of view in cognitive sociology. In P. Strydom (Ed.), *New trends in cognitive sociology*, Special issue, *European Journal of Sociological Theory, 10*, 465–480.

Bouvier, A. (2008). La théorie sociologique générale comme système hiérarchisé de modèles de portée intermédiaire. *Revue européenne de sciences sociales, t., 46*(140), 87–106.

Bouvier, A. (2011a). Individualism, collective agency and the 'micro-macro relation'. Chapter 8 of the *Handbook of philosophy of social science* (pp. 198–215). California: Sage Publications.

Bouvier, A. (2011b). Review of Peter Hedström and Peter Bearman (dir), *The Oxford Handbook of Analytical Sociology*, Oxford University Press, 2009, in *Revue française de sociologie* (pp. 373–379).

Bouvier, A. (2016). Analytical sociology, argumentation and rhetoric. Large scale social phenomena significantly influenced by apparently innocuous rhetorical devices. In: D. Mohammed & M. Lewinski, *Argumentation and reasoned action, proceedings of the first European conference on argumentation, Lisbon 2015* (pp. 291–299), Vol. II, Chap. 12.

Coleman, J. (1990). *Foundations of social theory*. Harvard: H.U.P.

Cook, K. S., & Gerbasi, A. (2009). *Trust*. In P. Hedström & P. Bearman (pp. 218–241).

Demeulenaere, P. (2011). Causal regularities, action and explanation. In P. Demeulenaere (Ed.), *Analytical sociology and social mechanisms*. Cambridge: C.U.P.

Di Iorio, F. (2015). *Cognitive autonomy and methodological individualism: The interpretative foundations of social life*. Berlin and New York: Springer.

Di Nuoscio, E. (1996). *Le ragioni degli individui. L'individualismo metodologico di Raymond Boudon*. Messina: Rubbettino.

Dray, W. H. (1957). *Laws and explanation in history*. Oxford: O.U.P.

Elster, J. (1979). *Ulysses and the Sirens. Studies in rationality and irrationality*. Cambridge: C.U.P.

Elster, J. (1983). *Sour grapes. Studies in the subversion of rationality*. Cambridge: C.U.P.

Elster, J. (1989). *The cement of society. A study of social order*. Cambridge: C.U.P.

Elster, J. (1990). *Psychologie politique. Veyne, Zinoviev, Tocqueville*, Paris, Ed. Minuit (English Trans. 1993).

Elster, J. (1998). A plea for mechanisms. In P. Hedström, & R. Swedberg, (dir.), 1998a, *Social mechanisms. An analytical approach to social theory* (pp. 45–73). Cambridge: C.U.P.

Elster, J. (2005). *Ulysses unbound. Studies in rationality, precommitment and constraints.* Cambridge: C.U.P.

Elster, J. (2007). *Explaining social behavior. More nuts and bolts for the social sciences.* Cambridge: C.U.P.

Festinger, L. (1957). *A theory of cognitive dissonance.* Evanston (Ill.): Row & Peterson.

Fleetwood, S. (1995). *Hayek's political economy: The socio-economics of order.* London: Routledge.

Freedman, J. L., & Frazer, J. C. (1966). Compliance without pressure: The foot-in-the-door technique. *Journal of Personality and Social Psychology, 4,* 195–202.

Gambetta, D. (Ed.). (1988). *Trust: Making and breaking of cooperative relationships.* Oxford: Blackwell Publishers.

Gambetta, D. (2009). Signaling. In Hedström & Bearman, 2009a, Chap. 8 (pp. 168–194).

Gardiner, P. (1952). *The nature of historical explanation.* Oxford: O.U.P.

Gilbert, M. (1989). *On social facts.* Princeton: Princeton University Press.

Gray, J. (1984). *Hayek on liberty.* London: Routledge.

Hayek, F. A. (1952). *The sensory order. An inquiry into the foundations of theoretical psychology.* Chicago: Chicago Univ. Press.

Hayek, F. A. (1962). Rules, perception, and intelligibility. *Proceedings of the British Academy, XL* (VIII), 321–344.

Hedström, P., & Bearman, P. (2009a). *Handbook of analytical sociology.* Oxford: O.U.P.

Hedström, P., & Bearman, P. (2009b). What is analytical sociology all about? An introductory essay. In P. Hedström, & P. Bearman, 2009a, Chap. 3 (pp. 25–47).

Hedström, P., & Swedberg. (1998). Social mechanisms: An introductory essay. In Hedström & Swedberg (Eds.) (1998), *Social mechanisms: An analytical approach to social theory.* Cambridge: C.U.P.

Hedström, P., Swedberg, R., & Udehn, L. (1998). Popper's situational analysis and contemporary sociology. *Philosophy of the Social Sciences, 28*(3), 339–364.

Hedström, P., & Ylikoski, P. (2014). Analytical sociology and rational-choice theory. In Manzo, 2014 (pp. 57–70).

Heider, F. (1958). *The psychology of interpersonal relations.* New York: Wiley.

Hempel, C. (1942). The function of general laws in history. In M. Martin et L. McIntyre, 1994 (pp. 43–54).

Hempel, C. (1962). Explanation in science and in history. In R. G. Colodny (Ed.), *Frontiers of science and philosophy* (pp. 7–33). Pittsburgh: University of Pittsburgh Press.

Horne, A. (1977). *A savage war of peace: Algeria, 1954–1962.* London: Macmillan (French Trans.: *1980, Une histoire de la guerre d'Algérie.* Paris: Alin Michel).

Kiesler, C. A. (1971). *The psychology of commitment. Experiments linking behaviour to belief.* New York: Academi Press.

Koyré, A. (1971). *Mystiques, spirituels, alchimistes du XVI siècle allemand.* Paris: Gallimard.

Manzo, G. (Ed.). (2014a). *Analytical sociology. Actions and Networks.* Chichester: Wiley.

Manzo, G. (2014b). Data, generative models, and mechanisms: More on the principles of analytical sociology. In Manzo, 2014a, (pp. 4–52).

Merton, R. (1967). On sociological theories of the middle-range. In *On theoretical sociology.* New York: The Free Press.

Milgram, S. (1963). Behavioral study of obedience. *Journal of Abnormal and Social Psychology, 67*(4), 371–378.

Nadeau, R. (1993). A bad argument for good reasons. *International Studies in the Philosophy of Science, 7*(1), 69–73.

Opp, K. (2013). What is analytical sociology? Strengths and weaknesses of a new sociological research program. *Social Science Information, 52*(3), 329–3600.

Peter, F., & Spiekermann, K. (2011). Rules, norms and commitments (Chap. 9). In *Handbook of philosophy of social science* (pp. 217–239). Sage Publications.

Pettit, P. (2005). Construing sen on commitments. *Economics and Philosophy, 21,* 15–32.

Peyrefitte, A. (1997). *De Gaulle et le Québec.* Montréal: Les Publications du Québec.

Popper, K. (1957). *The poverty of historicism*. London: Routledge.

Romdenh-Romluc, K. (2013). First-person awareness of intentions and immunity to error through misidentification. *International Journal of Philosophical Studies, 21*, 493–514.

Schelling, T. (1978a). *Micromotives and microbehaviors*. New York: Norton.

Schelling, T. (1978b). *Egonomics or the art of self-management. American Economic Review, Norton, 68*(2), 290–294.

Sen, A. (1977). Rational fools. A critique of the behavioral foundations of economic theory. *Philosophy & Public Affairs, 6*(4), 317–344.

Sperber, D. (1996). *Explaining culture: A naturalistic approach*. Oxford: Blackwell.

Sperber, D. (2011). A naturalistic ontology for mechanistic explanations in the social sciences. In P. Demeulenaere (Ed.), *Analytical sociology and social mechanisms*. Cambridge: C.U. P.

Thompson, D. C. (1990). *De Gaulle et le Québec*. Montréal: Éditions du Trécarré.

Tuomela, R. (2002). *The philosophy of social practices. A collective acceptance view*. Cambridge: C.U.P.

Tversky, A., Kahneman, D., & Slovic, P. (1980). *Judgment under uncertainty. Heuristics and biases*. New York: Cambridge University Press.

van Reybrouck, D. (2014[2010]). *Congo, the epic story of a people*. Harper Collins Publisher.

von Mises, R. (1949). *Human action. A treatise on human action*. New Haven: Yale University.

Zarefsky, D. (1990). *Lincoln, douglas, and slavery: In the crucible of public debate*. Chicago: University of Chicago Press.

Zarefsky, D. (1998). Four senses of rhetorical history. In K.-J. Turner (Ed.), *Doing rhetorical history: Concepts and cases*. Tuscaloosa: University of Alabama Press.

Zarefsky, D. (Ed.). (2014). *Rhetorical perspectives on argumentation. Selected essays*. Dordrecht: Springer.

Alban Bouvier PhD Sorbonne, Paris. Professor at Aix-Marseille University and Institut Jean Nicod (CNRS/ENS/EHESS, Paris). Bouvier has published papers in *Philosophy of the Social Sciences, Episteme. A Journal of Social Epistemology, The European Journal of Sociological Theory, Archives de Philosophie, Revue Internationale de Philosophie, L'Année sociologique, La Revue Française de Sociologie*. He has also published and edited or jointly edited several books, including: *L'argumentation philosophique. Etude de sociologie cognitive*, (Paris, Presses Universitaires de France), 1995 ; *Philosophie des sciences sociales. Un point de vue argumentativiste en sciences sociales* (Paris, Presses Universitaires de France), 1999 ; (ed.), *Pareto aujourd'hui* (Paris, Presses Universitaires de France), 1999 ; (co-ed. with R. Boudon and F. Chazel), *Cognition et sciences sociales*, 1997, (Paris, Presses Universitaires de France) ; (co-ed. with Bernard Conein), *L'épistémologie sociale*, 2007 (Paris, Editions de l'EHESS) ; (co-ed. with Raphaël Künstler), *Croire ou accepter*, 2016, Paris, Herman.

Chapter 4
On the Explanation of Human Action: "Good Reasons", Critical Rationalism and Argumentation Theory

Enzo Di Nuoscio

Abstract Raymond Boudon's main contribution to contemporary methodological debates is his definition of methodological individualism as a logic of explanation. In his opinion, social phenomena must be explained in terms of intentional and particularly unintentional consequences of the aggregation of human actions dictated by "good reasons". Boudon's approach can be divided into two complementary stages: the explanation of action, and the explanation of its consequences.

Raymond Boudon's main contribution to contemporary methodological debates is his definition of methodological individualism as a logic of explanation. In his opinion, social phenomena must be explained in terms of intentional and particularly unintentional consequences of the aggregation of human actions dictated by "good reasons". Boudon's approach can be divided into two complementary stages: the explanation of action, and the explanation of its consequences. Using a sophisticated theory of rationality, Boudon proposes to explain human action in terms of "good reasons"; while using his theory of unintended consequences, he proposes to explain those social events that extend beyond the individual intentions as the unwanted products of the spontaneous aggregation of human actions.[1] In other words, for Boudon, to explain a phenomenon social sciences need first to explain why individuals have acted in a certain way, and, then how individual actions have generated the phenomenon under investigation. Boudon's methodology focuses on two approaches that represent two facets of the same logic of explanation: i.e. the *situational analysis*, which is used to explain the causes of action, and the *invisible hand explanation*, which is used to explain the (unintended) consequences of action.

---

[1] Boudon (2007, pp. 37 ss). Also Dubois (2000, pp. 21ff), Hamlin (2001, pp. 67ff) and Di Nuoscio (1996, pp. 120ff).

E. Di Nuoscio (✉)
University of Molise, Campobasso, Italy
e-mail: enuoscio@luiss.it

E. Di Nuoscio
LUISS Guido Carli University, Rome, Italy

© Springer International Publishing AG, part of Springer Nature 2018
G. Bronner and F. Di Iorio (eds.), *The Mystery of Rationality*,
https://doi.org/10.1007/978-3-319-94028-1_4

37

Boudon's theory of rationality is a particularly useful instrument for identifying the "good reasons" to believe in certain types of ideas—such as false beliefs, ideologies, ethical norms and principles of justice—traditionally interpreted according to collectivist canons, i.e. in terms of sociological determinism. With his proposal for an individualistic explanation of such ideas, Boudon extends the notion of rationality to this important area of knowledge. Moreover, he insists on the need to identify sufficiently "adequate" reasons to explain the adherence to these ideas: what he calls the agent's "good reasons".[2]

In this chapter, I attempt to highlight the methodological potential of Boudon's approach, comparing it with some of the most important contributions to twentieth-century social philosophy: i.e. Ch. Perelman's theory of argumentation, K. Popper's and H. Albert's critical rationalism, L. von Mises' theory of rationality, and W. Dray's philosophy of action.

Good Reasons and Argumentation Theory

According to Boudon, *actions* and *beliefs* are the result of a single cognitive process through which the individual *simplifies* and ascribes *significance* to a kaleidoscopic infinity *void of significance*. The agent *acts* and *believes* on the basis of *good reasons*, i.e. by means of a strategy that is *subjectively* rational, based on a rationality that is both *bounded* and *situated*. For Boudon, there is thus no difference between rationality of action and rationality of beliefs. Here we do not have two *discontinuous* genres but two manifestations of the same *genus* represented by the rational calculus as a result of which each single actor decides to believe and act. Moreover, Boudon argues that, regarding beliefs, those that are false or normative are not less rational than those that are commonly regarded as true and/or scientific. In many cases—Boudon states—the mechanisms that produce unfounded beliefs are nothing more than ordinary mechanisms of thought.[3] As a consequence, the process of adherence to a belief is independent from its contents.[4]

In this manner, Boudon 'destroys' all the possible barriers that may exist between actions and beliefs, as well as between positive and normative beliefs. Actions, scientific theories, magical beliefs, moral norms and ideologies are all products of similar decisional mechanisms. To explain these mechanisms, it must be considered that, from his/her *position*, the individual acts upon or adheres to an idea on the basis of motivations that are, according to his/her personal standpoint, *good reasons*. Moreover, the individual develops his/her view using some fundamentally implicit a priori. Clearly, this does not mean that there is no difference among science, magic, ideology and morality; merely that the individual, in developing magical or normative beliefs, does not use alternative cognitive mechanisms.

---

[2]Boudon (2010, pp. 55ff).

[3]Boudon (1995, p. 102).

[4]Boudon (2011, pp. 33 e ss).

Given the above, Boudon argues that to explain an action or any sort of belief we must follow a single, explanatory logic: we need to conjecturally reconstruct the situation as perceived by the agent and retrace the reasons that have induced the agent to act as s/he did or believe what s/he believed. In the process, we must be careful to make explicit the various a priori assumptions implicit in the agent's reasoning.

Both actions and beliefs are the result of a *subjectively rational calculus*. This calculus is comprised of both implicit and explicit argumentations. According to Boudon, what actions and beliefs have in common is their *argumentative foundation*. The reconstruction of the agent's argumentations which are often implicit, i.e. not made public by the agent, allows the methodological individualist to explain human behaviour.[5] Considering argumentations as the rational foundation not just of actions and true beliefs, but also of false or normative beliefs, Boudon detaches himself from the narrow conception of argumentation rooted in positivism. Positivism "has induced us to consider as manifest the idea that *only objectively valid* reasons can have a causal influence on the convictions of a social subject".[6] Against this view, Boudon asserts that the progress of ordinary and scientific knowledge (as well as the evolution of moral ideas) occurs chiefly on the basis of reasons that are originally merely subjectively valid.

According to Boudon, there cannot exist, as the positivists maintained, a dualism between scientific and non-scientific arguments, because both are of one and the same nature. If one considers, as neo-positivists did, only the argumentations based on *objectively* valid reasons to be sensible and also that *subjectively* valid reasons have no causal power, the field of rationality would be reduced to such an extent that many beliefs would be considered irrational or even, as some neo-positivists have asserted, *nonsensical*. "There is no correlation", Boudon observes, "between the field of application of the argumentative approach and the solidity of the reasons that per se it sets in motion. If we are correct then there is no reason to accept the positivist idea according to which, in an extra-scientific context—e.g., in the field of morality or in that of ordinary thought—the reasons should be considered as simple 'rationalizations' without any effect whatsoever on beliefs".[7]

By insisting that reasons are only subjectively valid, Boudon believes that actions and beliefs are rational even when they are the result of *non-demonstrative* argumentations. He therefore develops a view similar to that of Chäim Perelman, the Polish philosopher author of one of the most important contributions to the contemporary theory of argumentation, which is based on subjective and bounded rationality. To reduce the sphere of irrational behaviours and beliefs, Perelman enlarges the notion of rationality by proposing a conception of rational argumentation that transcends Cartesian *more geometrico* reasoning. Perelman and Olbrechts-Tyteca argue that if we were to consider as misleading reasoning any argumentation of this

---

[5]Alban Bouvier has widely analyzed the heuristic utility of the argumentative approach to enhance the explanatory power of methodological individualism: Bouvier (1995, pp. 113ff).

[6]Boudon (1995, p. 91, our emphasis).

[7]*Op. cit.*, p. 92.

type [non-demonstrable], the insufficiency of logic-experimental proof would clear the way, in all basic fields of human life, for suggestion and violence. By pretending that all that is not objectively and indisputably valid depends on the subjective and arbitrary, we would generate an insurmountable abyss between historical knowledge —believed to be the only rational type of knowledge—and the motivations of action which would be completely irrational.[8]

It could be objected to this approach that, since the *practical reasoning,* as well as the *theoretical reasoning,* which must be conceptually 'reconstructed' by the methodological individualist, are often implicit, it is incorrect to speak of argumentations. It could be argued that no argumentations are developed if the agent does not share his/her reasons with other people. Moreover, even when the agent makes the reasons behind his/her behaviour explicit, these reasons could be justifications (i.e. Paretian a posteriori *derivations*), which means that we could not regard these reasons as the real causes of his/her action or adherence to a belief. It is important to stress that the good reasons that the researcher must study and "reconstruct" are the reasons that the agent *really* had to act, not the motivations that the agent pretended to be his/her motivations to act. For Boudon, if the agent provided false information, his/her false information is for the researcher part of the information that the researcher needs to investigate so as to discover the agent's *real* reasons for acting.

Regarding the idea that the motivations of the agent are not real argumentations because they are implicit, one could object—following Perelman—that agreement with oneself is a particular case of agreement with others[9] and that, therefore, although it does not imply a public, deliberation with oneself is a particular case of argumentation.[10] When s/he decides, the individual is concerned with collecting all the arguments that manifest some value, without having to keep any of them tacit,

---

[8]Perelman and Olbrechts-Tyteca (1969, p. 404). In the "Preface" to the Italian version of this work, Norberto Bobbio writes the following: "Through the distinction between argumentation and demonstration, between conviction and persuasion, between logic *strictu senso* and rhetoric, between demonstrative reasoning that has value independently of the persons to which it is directed and persuasive reasoning that has value only in reference to a determined audience, the theory of argumentation was presented as an attempt to bring, once again, ethics into the domain of action, even if the reason in question was practical and different from the pure or, if you prefer, as the discovery of a field that for too long has remained unexplored after the triumph of mathematical rationalism between those occupied by the invincible force of reason and, on the opposite side, by the invincible reason of force [...]". The *new rhetoric* by Perelman is a theory about non-demonstrative rationality, which is extremely important because, as Bobbio explains, "where values are at stake, it does not matter if sublime or vulgar, demonstrative reason, the one that refers to logic in the strict sense, is important: all that is left is their inculcation (or oppression) or finding some 'good reasons' to support (or to confute) them. The rhetoric of argumentation is the methodical study of the good reasons that humans use to take and to discuss choices that implicate a reference to when they have refused to impose them through violence or to strip them away with psychological coercion, i.e., through imposition or indoctrination"; N. Bobbio, *Preface* to Perelman and Olbrechts-Tyteca (1986).

[9]C. Perelman and L. Olbrechts-Tyteca, *The New Rhetoric: A treatise on Argumentation*, cit., p. 38.

[10]*Op. cit.*, p. 38.

and to freely decide, after balancing pros and cons, on the solution that s/he believes to be the best.[11] The agent performs a comparative evaluation that can only be developed through argumentations which have the agent per se as an audience. It is evident that the agent manages to persuade him/herself to believe something through *reasoning*.

"Good Reasons" and Critical Discussion

Arguing that all beliefs are based on good reasons, i.e. on rational argumentations, does not necessarily mean accepting a *tyranny of opinion*, i.e. assuming that all opinions are equal and/or incommensurable. According to Boudon, the agent's good reasons must be revealed and critically discussed, making the implicit a priori of the argumentations that underlie the individual *deliberations* explicit and clear. Following this approach, Boudon criticizes relativist post-Popperian epistemologists such as Kuhn and Feyerabend. He reconstructs the good reasons which lead these epistemologists to conclusions that are relativistic. He then strongly criticizes their relativism, showing that some implicit assumptions of their argumentations are unacceptable. Boudon highlights how these philosophers of science start from good reasons but arrive at conclusions that he considers wrong in that they are based on unacceptable a priori assumptions.[12]

The rationality of good reasons can be thus regarded as a *critical rationality,* in that this rationality allows critical comparison of two or more ideas/theories, dissecting their respective reasons and argumentations. The result of this critical discussion, whatever it may be, must in turn be motivated by good reasons. Clearly, the critical comparison of positive theories is different from the critical comparison of normative theories, i.e. of theories based on value choices. However, the critical method is the same: reconstructing the reasons and the argumentations underlying each belief, before critically comparing these reasons and argumentations. According to this approach, the kind of critical evaluation that the individual performs at the *intra-subjective* level when s/he decides must be regarded as identical to the critical evaluation performed at the *inter-subjective* level.

Boudon's methodological theory of critical discussion, which supports the possibility of comparing subjectively rational argumentations, is particularly useful for the study of normative beliefs. If the genesis of these beliefs is interpreted as the result, both intentional and unintentional, of cognitive procedures that are subjectively rational, it must be concluded that these procedures can be discussed and compared. If, on the contrary, one attempts to explain them through argumentations

---

[11]*Op. cit.*, pp. 37–38. In these cases, Perelman and Olbrechts-Tyteca contend, the secrecy of intimate deliberation seems to warrant its sincerity and value; *op. cit.*, p. 38.

[12]Di Nuoscio (2004a, pp. 66ff).

that are *demonstrative*, there would be no room for rational discussion and no possibility of comparison: we would be in the domain of the irrational.[13]

A critical comparison of ideologies or moral theories is possible only if we admit that they may be, directly or indirectly, supported by typical factual argumentations. Because of Hume's Law, ethical standpoints cannot be logically derived from factual judgements; rather, according to the argumentation theory, ethical standpoints can be supported by empirical truths. The individualistic explanation of normative beliefs is possible only by avoiding an *extensive* interpretation of *Hume's Law*, i.e. rejecting the idea that no relationship between ethics and cognitive propositions can be built.[14] It is true that propositions concerning individual values cannot be *rationally demonstrated*. However, they can be *argued and discussed* by making reference to judgements of fact.[15] As argued by Perelman before Boudon, the practice and theory of argumentation are correlated with a *critical rationalism* which transcends the reality judgement/value judgement dualism, and which strongly relates to the personality of both the scientist and of the philosopher; where these latter are responsible for their decisions in both the field of knowledge and in that of action.[16]

Moral views, just like ideologies, are the result of competition between diverse reasons and argumentations. In consequence, beliefs of this kind are *without truth* for they cannot be rationally founded; but they are not *without reasons*, because they can be argued and discussed; the freer the discussion, the more authentic the *rationality* of beliefs will be.[17] According to Perelman, the existence of an argumentative process which is not *binding* or *arbitrary* gives meaning to human freedom, the condition for the exercise of a reasonable choice. If freedom were only necessary adherence to a natural order given *ex ante*, it would exclude every possibility of choice: if the exercise of freedom were not founded on some reasons, each choice would be irrational and would reduce to an arbitrary decision which would act in an intellectual vacuum. Thanks to the possibility of an argumentation which furnishes reasons, but not binding ones, it is possible to escape the dilemma:

---

[13]C. Perelman, L. Olbrechts-Tyteca, *The New Rhetoric: A Treatise on Argumentation*, cit., p. 461.

[14]*Op. cit.*, pp. 462ff.

[15]This is the case, for example, of the choice for democracy, which is founded on a value scale (the protection and the development of individuals) that can only be effective in a regime that guarantees the respect of individual freedoms. The choice of democracy can be defended by stating that the competition among free people allows the greatest economic welfare and the greatest growth of knowledge. Hans Albert argued that, although it is true that we cannot infer a value judgement from an empirical statement, it is also true that the progress of our empirical knowledge can show that certain of our value judgements are *incompatible* with some of our other ethical convictions; Albert (1968), Italian translation, il Mulino, Bologna, 1973, pp. 99–100.

[16]C. Perelman, L. Olbrechts-Tyteca, *The New Rhetoric: A Treatise on Argumentation*, cit., p. 538, emphasis added.

[17]According to Perelman, like science, also philosophy, morality and law take their rationality from the system of argumentation, from the good reasons that we can advance in favour and against any thesis in question; Perelman (1982, p. 170).

adherence to a truth objectively and universally valid, or recourse to undue influence and violence in order to have one's decisions and opinions accepted.[18]

Perelman's *critical rationalism*, which is implicit in Boudon's theory of good reasons, presents some similarities with critical rationalism as understood by Popper and Albert. Boudon introduces a third category between *foundationalism* and *subjectivism*. Indeed, Boudon's good reasons cannot be a *fundamentum inconcussum* (i.e. a firm foundation) for arguments, as good reasons are only subjectively valid. Quite often, what Boudon calls "good reasons" are not reasons based on a demonstrative or Cartesian rationality even if they are developed through *intersubjective* argumentations; accordingly, good reasons do not presuppose any universal agreement. In Boudon's opinion, what makes the explanation of an action in terms of methodological individualism possible is the fact that the agent's reasons are understandable and can be subject to criticism even by people who belong to cultural environments different from that of the agent. Although based on subjective rationality, Boudon's good reasons thus have a trans-cultural causal relevance. Even if there are temporal, spatial and cultural differences between the agent and the researcher, the latter can reconstruct the causal relationship between the agent's action and its good reasons through what Popper has called "objective comprehension"[19]—i.e. via a conceptual reconstruction of the situation through the conjectures and refutations method.

On the Complementarity Between Mises' "Praxeology" and Boudon's Cognitive Rationality

As we have seen, for Boudon the rationality of actions and beliefs can be understood via a complex process which requires the reconstruction of the situation as perceived by the agent and the disclosure of the implicit a priori underlying the agent's decision strategies. This process enables the researcher to reconstruct the

---

[18]*Ibid.* In the *Discourse on Method* Descartes wrote as follows: "considering that of all those who have hitherto searched for truths in the sciences, only the mathematicians have been able to find some demonstrations, namely some *certain and evident reasons*, I had no doubt that I should start from the same things that they had analyzed; even if I did not hope to derive any utility from them, if not that they would accustom my mind to nourishing itself on truths and not being happy with *false reasons*" (p. 72, emphasis added). If the only valid reasons are those that are "certain and manifest...considering then how many different opinions on a same manifestation the scholars are able to sustain, without more than one that can be true, I believed almost false everything that was only similar to truth" (p. 62); Descartes (1998). Descartes is just one—albeit one of the most important—representatives of a long tradition in Western thought that has tried to found scientific, metaphysical, ethical and political *ideas* on a firm *fundamentum inconcussum*. Various expressions have been used to refer to this firm foundation: e.g. 'self-evident principles' (rationalistic tradition), sensations (empiricists), the Reason (Enlightenment thinkers), 'facts' (positivists), 'economic structure' (Marxists) and 'verifiability principle' (neo-positivists).
[19]Popper (1972, pp. 226 ff).

agent's decision-making process that is subjectively rational. Boudon underlines that it could occur that, notwithstanding his/her efforts, the researcher fails to reconstruct the reasons which, on the basis of his/her information about the agent and the situation, s/he can consider the agent's *good reasons*, i.e. the causes of the agent's action or belief. In these cases, the researcher is legitimated to conclude that the agent has acted irrationally, arguing that the situational analysis did not reveal the existence of good reasons. For Boudon, the researcher needs to motivate his/her inference of irrationality.

Boudon's theory of subjective rationality is seemingly inconsistent with Mises' praxeology. Praxeology is based on the postulate that "human action is always necessarily rational".[20] According to praxeology, this postulate is correct for two reasons: (i) action is always a choice between different means to an end; and (ii) the agent always selects the means that, according to his/her knowledge, are best suited to the intended purpose. According to Mises, human action is always rational.[21]

Mises' approach seems to conflict with Boudon's on what may be considered the core of any variant of methodological individualism, i.e. the theory of rational action. On the one hand, according to Mises' praxeology, the "fundamental categories of action" (i.e. actions, preferences, economization, selection of means the achievement of a goal, etc.) have the status of a priori logical and mathematical truths, without reference to any experience whatsoever.[22] On the other, according to Boudon's cognitive perspective, rationality, when it is discovered, is the result of a long investigative process. Therefore, it can only be discovered a posteriori. It is related to reasons that can only be reconstructed *ex post*, not predicted *ex ante*.

However, the contrast between Mises and Boudon is probably only superficial.[23] In fact, the two scholars seem to adopt two diverse analytical standpoints. They conceive rationality in different but not necessarily inconsistent ways. Boudon's rationality refers to the *sociological nature of reasons*, while Mises' praxeology concerns the *logical structure of action*. For Boudon, action is rational if the reasons behind it are rational, i.e. if empirical research allows us to find reasons for acting that may be considered, in a particular social environment, causally adequate in producing that particular action. According to Boudon, the entire logic of individualistic explanation must be based on this *interpretative* approach as well as on the analyses the unintentional consequences produced by the aggregation of single actions. Boudon's rationality can be regarded as a sort of ideal model that guides the researcher in investigating the problematical situation as perceived by the agent before acting. In Boudon's opinion, the most efficacious way in which to explain action is to reconstruct empirically the good reasons for acting developed by the individual in a given environment.

---

[20]von Mises (1949, p. 22).

[21]von Mises (2011, p. 153).

[22]von Mises (1976, p. 41).

[23]I developed this thesis in Di Nuoscio (2009, pp. 175–194), See also Di Iorio (2015, p. 121 ff.).

By contrast, Mises' praxeology is an a priori general theory of human action which does not focus on the psychological and sociological motivations of action,[24] but on the "self-evident"[25] principles that represent the *analytical* attributes of action. The first among these attributes is the idea that every action necessarily seeks to remove a certain uneasiness by pursuing an end. The agent—*by definition*—pursues his/her own ends, selecting the means that, to the best of his/her knowledge, are the most appropriate. According to Mises, the task of praxeology is to study means, not ends. Furthermore, Mises argues that the researcher should not comment upon the agent's choice of means, in the sense that the researcher should use a descriptive concept of rationality rather than a normative one.[26] Mises' concept of rationality derives from *reason*, not from *experience*.

Boudon—who unfortunately never took a stance on Mises' praxeology—would certainly agree with Mises that, since human action seeks to remove a certain uneasiness through a means/ends calculation, human action is *in this sense* always rational. However, being a comprehensive sociologist, Boudon was mainly interested in the situational analysis that can be developed by reconstructing the agent's reasons for acting. From Boudon's viewpoint, it is important to stress that the situational analysis can work only if we assume that not all the reasons which the agent could have for acting can be used to explain his/her action. The researcher should not stop his/her analysis at the most superficial and simple reasons the agent could have for acting, but should go deeper providing a detailed analysis of the situation.

Accordingly, Boudon conceives the rationality of actions and beliefs in a different way than that of Mises. Boudon does not refer to the generic relationship between means and ends, but to the relationship between reasons and actions. As a result, for these two scholars, also irrationality has a different meaning.

---

[24] According to Mises, praxeology and catallaxy do not deal with the motives and the ultimate ends of action, although they are means applied to achieve a given end (*Human Action*, cit., p 16); moreover, the principles of praxeology and economics are valid for all human actions irrespective of the motivations, causes and ends underlying them (*op. cit.*, p. 21). "Everything that we say about action"—Mises contends—"is independent of the motives that cause it and of the goals toward which it strives in the individual case. It makes no difference whether action springs from altruistic or from egoistic motives, from a noble or from a base disposition; whether it is directed toward the attainment of materialistic or idealistic ends; whether it arises from exhaustive and painstaking deliberation or follows fleeting impulses or passions. The laws of catallactics that economics expounds are valid for every exchange regardless of whether those involved in it have acted wisely or unwisely or whether they were actuated by economic or noneconomic motives. The causes of action and the goals toward it strives are data that are needed for the theory of action: upon their concrete configuration depends the course of action taken in the individual case, but the nature of action as such is not thereby affected" (*Epistemological Problem of Economics* cit., p. 57). "Praxeology is a theoretical and systematic, not a historical, science. Its scope is human action as such, irrespective of all environmental, accidental, and individual circumstances of the concrete acts. Its cognition is purely formal and general without reference to the material content and the particular features of the actual case": (*Human Action*, cit., p. 31).

[25] von Mises (1962, p. 4).

[26] L. von Mises, *Human Action*, cit., p. 15.

Mises assumes that condemning an action as irrational always has its roots in an evaluation made on the basis of a value scale which is different from ours.[27] Conversely, Boudon thinks that irrational actions are actions that are not dictated by good reasons because the researcher could not find reasons that can be considered adequate to explain the action empirically. Since Boudon's concept of rationality refers to the *sociological nature of reasons*, it is not just a logical characteristic of action, but also a socially influenced characteristic of action. This is because, for Boudon, the causal adequacy of reasons can only be evaluated on the basis of the social practices prevailing in a certain cultural environment, i.e., as argued by Weber, "according to our habitual modes of thought and feeling".[28]

The Epistemological Status of Rationality

Clarifying the distinction between the *structure of action* and the *nature of reasons*, i.e. between Boudon' s concept of rationality and Mises's, is useful to clarify the status of rationality from a philosophical viewpoint.

Trying to provide a definition of rationality, Popper maintained that rationality cannot be regarded as a *psychological thesis* according to which people always acts rationally, in all or in the majority of cases.[29] Nor can rationality be considered a true "explanatory theory" or "testable hypothesis", because these characteristics can be ascribed only to the situational analysis, not to the rationality principle.[30] However, even though it is not empirical, for Popper nor can rationality be considered a priori, for it is "clearly false"[31] that we do not always act in tune with the rationality principle.[32]

Arguing that there are inadequate behaviours produced by the agent's interpretation of the situation, Popper clearly denies that the rationality principle is a priori, i.e. characterized by universal validity. However, he also argues that the rationality principle is empirically untestable. As a consequence, by arguing that the rationality principle is false, Popper contradicts himself. He tries to make sense of this contradiction by arguing that the rationality principle must nonetheless be

---

[27]Mises argued thus: if we wanted to distinguish rational action from irrational action, we would not only become judges of the value scales of our fellow men but we would also be saying that our knowledge is the only correct, objective standard of knowledge. Anyone who states that irrationality plays a role in human action simply states that his fellows behave in a way that he does consider as correct. If we do not desire to judge the ends and value scales of others, to claim our omniscience, the statement that 'he behaves irrationally' is meaningless because it is incompatible with the concept of action.; L. von Mises, *Epistemological Problem of Economics*, cit., pp. 56–57.

[28]Weber (2013, vol. I, p. 10).

[29]Popper (1985, p. 359).

[30]*Op. cit.*, p. 360.

[31]*Op. cit.*, p. 366.

[32]*Ibidem.*

considered an integral part of every, or almost every, controllable social theory.[33] In his opinion, if a theory is falsified, it is a good practice, a correct methodological expedient to abstain from blaming rationality for the failure of our theory.[34]

Thus, what is falsifiable is the conjecture about how the situational analysis is linked to the principle of rationality. This principle must not be regarded as responsible for the falsification of a theory. Although the rationality principle does not have universal validity, Popper recommends using it because it represents a "good approximation of the truth".[35] Hence, if compared to other methods, it seems to yield explanatory hypotheses—that is, conjectural situational models—which are easier to control.[36]

Popper disagrees with Mises's praxeology in that he admits the possibility of irrational actions i.e., of actions that are not adequate for the situation. Analysing the difference between rationality and irrationality of action, Popper seems to refer, like Boudon, to the nature of reasons. According to Popper, an action is rational if the researcher, after having tried to reconstruct the situation as perceived by the agent, concludes that, in the same situation and having the same information, s/he would have behaved in the same way. For Popper, the action must be evaluated on the basis of *shareable* reasons—thus, according to the rationality principle. However, Boudon partly criticized Popper for the way in which he distinguished between the rationality and irrationality of actions because Popper did not make any difference between *reasons* and *good reasons*.

Popper's unsatisfactory definition of the epistemological status of the rationality principle likely depends on the fact that he failed to draw a clear distinction between action and reasons—a distinction whose importance is evident if one compares Mises' praxeology, on the one hand, and Boudon's and Popper's subjective concept of rationality on the other. With regard to action, the rationality principle is a priori and universally valid because—as stressed by Mises—every action is *always* and *necessarily* directed at pursuing a goal through the use of those means that the agent considers the most appropriate.

However, for comprehensive sociologists such as Boudon or Weber—and more generally for any author who, like Popper, insists on the heuristic utility of *situational analysis*—what is chiefly at stake is the reconstruction of the problematic situation as perceived by the agent to get the greatest amount of information about his/her action. Only on the basis of such a reconstruction, which needs to be empirically tested using the situational information gathered, can the researcher conclude that an action is rational, i.e. that s/he would have behaved in the same fashion as the agent did having the same information. In other words, the researcher needs to make an *evaluation*. S/he must judge if, considering the agent's perception

---

[33]*Op. cit.*, p. 362.

[34]*Ibidem.*

[35]*Ibidem.*

[36]*Op. cit.*, p. 364.

of the situation, the agent's reasons for acting can be considered shareable, i.e. generally acceptable and appropriate given the agent's goal.

On this view, the irrationality of action is neither a characteristic of the logic of action, nor an empirical truth, but an *evaluation* that the researcher makes by conjecturally and empirically reconstructing the agent's situation. This evaluation can only be made by considering the *cultural* background of the agent, i.e. the set of beliefs, habits and Weberian "rules of experience" that are dominant in the agent's environment. If referred to what I called above the *sociological nature of reasons*, the principle of rationality can be interpreted as a *normative principle* that establishes how individuals *should* behave in a given situation. Interpreted in these normative terms, the principle of rationality cannot be described as the principle, according to which, to use Popper's words, individuals do not always "behave in a manner appropriate to the situation". Understood in normative terms, the principle of rationality is, as stressed by William Dray, the principle according to which if we are in a situation of the C-type, the right thing to do (rational, appropriate, adequate) is to do x.[37]

If conceived in normative terms, the rationality principle concerns the motivations for acting. As a consequence, it is not falsified by the existence (admitted by Popper and especially by Boudon) of irrational actions. Naturally, a normative evaluation is possible only if both the observer and the observed, even if they belong to very different cultural environments, have some inferential principles in common; this represents the *conditio sine qua non* that makes possible the "reasoning by analogy" called *Verstehen*.[38]

Boudon's "Good Reasons" and Dray's "Principle of Action"

In *Law and Explanation in History*, which is one of the most influential contributions to the explanation of action, William Dray proposes an explanatory model similar to Boudon's. According to Dray, an action can be explained through a reconstruction of the *calculation* made by the agent concerning the means to adopt for the desired ends, in light of the circumstances that surround him/her. To explain action we must know what considerations convinced him/her to act in the way

---

[37]Dray (1957, p. 163).

[38]Robert Nozick has argued as follows: The inferences concerning some actor on the basis of *Verstehen* depend on the fact that you use your imagination in his/her place, that you see him/her as you see yourself. It is a form of reasoning by analogy, and to this reasoning we have always granted a certain role played by inductive logic and by the empirically backed theory. The *Verstehen* is a form of reasoning by specific analogy, where the thing the other is analogous in is myself. From this we infer that he/she is behaving as we would in that same situation; we have in mind a partially defined subjective situation; Nozick (1981, p. 636).

he/she did.[39] According to Dray, insofar as we state that an action is indeed an intentional action—it does not matter at which level of conscious deliberation— there is a calculation that we could build for it: the one that the agent would have passed through if he/she had the time, if he/she had seen what to do in an instant, if he/she had been required to explain what he/she did after the action, etc. And it is by highlighting some calculations of this type that we explain action.[40] For Dray, the "rational explanation" of action[41] is thus a reconstruction *post factum* of the reasoning—which can be either explicit or tacit—on which the individual action is based. In other words, in Dray's decision, the historian and the sociologist must decide—on the basis of the information in their possession concerning the agent's situation—if the reasons that they have identified can be considered adequate to provoke the agent's action. The aim of the explanation is to demonstrate that what has been done *was the thing to do* given certain reasons—rather than the simple thing that one does in those circumstances, perhaps according to some laws.[42] Hence, the explanation of action presents a "crucially important characteristic": in it there is "an element of *evaluation* of what has been done; what we want to know, when we ask for an explanation of an action, is in what sense the action was actually *appropriate*".[43]

Naturally, as Dray makes clear, it is not necessary for the historian to demonstrate that the agent was right (in doing what he did); all that is needed is for the explanation to show that he/she had some reasons. But the element of evaluation remains for those that the historian states were the reasons of the agent that must really *be* the reasons (from the point of view of the agent). Once one accounts for all the principles, the beliefs and the peculiar intentions of the agent, to report what the agent *said* were his/her reasons would not be sufficient to give a rational explanation if the persuasiveness of those reported actions could not be appreciated by the historian.[44] In essence, to develop a rational explanation of an action it is necessary to find some *good* reasons, at least in the sense that, *if* the situation were as the agent saw it (whether or not, from our point of view, we agree to view it as such), then what has been done was the thing to do. The historian must be able to 'solve' the calculation of the agent.[45] Even when the latter made an error in examining the situation—that is, in the case that the agent *saw* the situation in a mistaken way—we can still explain it in a rational way considering that, taking his/ her wrong beliefs into account, the calculation can be made in a satisfactory fashion.[46]

---

[39]W. DRAY, *Law and Explanation in History*, cit., p. 125.

[40]*Op. cit.*, p. 126.

[41]*Op. cit.*, p. 127.

[42]*Ibidem.*

[43]*Ibidem.*

[44]*Op. cit.*, p. 132.

[45]*Ibidem.*

[46]*Op. cit.*, p. 133.

Like Boudon, Dray assumes that in order for the *reasons* to be considered *good* by the researcher, they must be somehow generalizable; otherwise there is the risk that the researcher makes ad hoc explanations. According to Boudon, good reasons "are universal in the sense that the majority of the subjects placed in our same position"[47] would have behaved in the same fashion. The "principle of action" proposed by Dray supports a similar view: "If *y* is a good reason for *A* to do *x*, then *y* would be a good reason for *anyone* sufficiently like *A* to do *x* in sufficiently similar conditions".[48] As Dray explains criticizing C.G. Hempel's nomological model of explanation, the explanatory adequacy of the individual's reasons must be generalizable through evaluation, not through an empirical law. In fact, if we find a negative case for a general empirical law, we must modify or reject the law since the law asserts that people *behave* in a certain way considering certain circumstances. But if we find a negative case for the type of general statement that one could extract from a rational explanation, this would not be necessarily falsified.[49]

Both for Dray and Boudon, the rational explanation in terms of evaluation is not based on some form of intuition, but on the empirical reconstruction of the situation. Regarding this point, their view is similar to Weber's, although Dray curiously does not cite Weber. Dray and Boudon do not conceive the comprehensive method in intuitive and metaphysical terms, but regard it as based on an explanatory model that is not, after all, very different from Popper's situational analysis. According to Dray, a rational explanation is not disconnected from the empirical analysis. It is inductively based on empirical evidence. For Dray, if a historian wants to understand Disraeli's behaviour, s/he cannot just ask himself/herself: 'What would I have done in his place?'; s/he also needs to read Disraeli's letters, dispatches, and so on. Dray stresses that understanding the rational meaning of human actions means developing 'scientific' explanations; it has nothing to do with pure imagination.[50]

The reconstruction of a rational calculation in order to explain human action can only be made by analysing the situation, i.e. by considering what Dray calls "data" that are "external" to the individual.[51] Therefore, for Dray, like for Boudon, the comprehensive method requires a situational analysis. As a consequence, not just Boudon but also Dray is close to Popper's view (this despite the difference between Dray and Popper regarding the role of the nomological explanation in the social sciences).[52]

---

[47]Boudon (1997, p. 385).

[48]W. Dray, *Law and Explanation in History*, cit., p. 133, our emphasis.

[49]*Op. cit.*, p. 134.

[50]*Op. cit.*, pp. 130–135.

[51]*Op. cit.*, p. 129.

[52]Popper's situational analysis is consistent with the covering law model and thus his theory of a unified method. Popper argues that causal and logical explanations are "perfectly compatible" (Popper 1957). However, he points out that the laws utilized in historical explanation "are usually devoid of interest, for the simple reason that they are, in the context, a-problematical" (Popper 1974, p. 120). Such laws, moreover, can be so trivial, so much a part of what we all know, that we do not need to specify them, and we rarely realize this. If we say that the cause of death is being

References

Albert, H. (1968). *Tractat über Kritische Vernuft*. Tübingen: Mohr.

Boudon, R. (1995). *Le juste et le vrai*. Paris: Fayard.

Boudon, R. (1997). *The art of self-persuasion: The social explanation of false beliefs (1990)*. Cambridge: Polity Press.

Boudon, R. (2007). *Essais sur la théorie générale de la rationalité*. Paris: PUF.

Boudon, R. (2010). *La sociologie comme science*. Paris: La Dècouverte.

Boudon, R. (2011). Ordinary rationality: The core of analytical sociology. In P. Demeulenaere (Ed.), *Analitacal sociology and social mechanism*. Cambridge: Cambridge University Press.

Bouvier, A. (1995). *L'argumentation philosophique*. Paris: PUF.

Descartes, R. (1998). *Discourse on method (1637)*. Cambridge (MA): Hackett Publishing Company.

Di Iorio, F. (2015). *Cognitive Autonomy and Methodological Individualism: The Interpretative Foundations of Social Life*. Berlin and New York: Springer.

Di Nuoscio, E., (1996). *Le ragioni delgi individui. L'individualismo metodologico di Raymond Boudon*. Soveria Mannelli: Rubbettino.

Di Nuoscio, E. (2004a). *La relatività del relativismo*. In R. Boudon, E., Di Nuoscio, & C. L. Hamlin (Eds.), *Spiegazione scientifica e relativismo culturale*. Roma: Luiss University Press.

---

burned alive, we do not need to recall the universal law that all living beings die if exposed to intense heat. But this law is implicit in our causal explanation (K. R. Popper, *The Poverty of Historicism*, cit., p. 129). According to Popper, nomological explanation is implicit in the historical explanation. In other words, meanings and motivations of actions which are of interest for the historian can be only recognized and used for the explanation because of the reference to laws. However, since these laws are so "trivial" "devoid of interest" and used in an a-problematical fashion by the historian, regarding social sciences the covering law model must be merged with a comprehensive or interpretative approach. Popper's situational analysis is precisely an interpretative approach (which is at odds with psychologism). As stressed by Popper in his autobiography, his situational analysis tries to "generalize the method of economic theory (marginal utility theory) so as to become applicable to other theoretical social sciences. In its most recent formulations, this method consists of the construction of a model of the social situation, including, in particular, the institutional situation in which the agent acts in such a manner as to explain the rationality (the zero character) of his/her action" (K. R. Popper, *Intellectual Autobiography*, cit., p. 121). Also in the *Poverty of Historicism* Popper argued that situational analysis consists in calculating "the deviation of the real behaviour of people from model behaviour" (K. R. Popper, *Poverty of Historicism*, cit., p. 126), that it is from that behaviour that we must expect on the basis of a 'pure logic of choice', as described in economic equations (op. cit., p. 127). Popper thus proposes an approach that is very similar to Weber's ideal-type of rational action. According to Weber, explaining actions means examining real actions as 'deviations' from ideal-types of action...built in a purely rational fashion with respect to purpose (M. Weber, *Economy and Society*, cit., p. 26). Popper and Weber agree that explaining an action means reconstructing the agent's rational calculus, i.e. the reasons and meaning of his/her action, on the basis of situational models. Since Popper criticizes both intuitionism and psychologism and defends what he calls "objective comprehension", his approach is not very different from Weber's *Verstehen*. While, according to Dray, the relationship between reasons and actions is reconstructed through generalizations which are emphatically developed, according to Weber, it is reconstructed through "rules of experiences". Weber also stresses that this relationship cannot be reconstructed through universal laws because, as we have seen, these rules of experience are falsified. Regarding the relationship between laws and the explanation of action and social macro-phenomena, see Di Nuoscio (2004b, cap. 5).

Di Nuoscio, E. (2004b). *Tucidide come Einstein. La spiegazione scientifica in storiografia.* Soveria Mannelli: Rubbettino.

Di Nuoscio, E. (2009). *The rationality of human action. Toward a Mises-Popper-Boudon model?* In M. Cherkaoui, & P. Hamilton (Eds.), *Raymond Boudon. A life in sociology,* Vol. II. Oxford: Bardwell Press.

Dray, W. (1957). *Law and explanation in history.* Oxford: Oxford University Press.

Dubois, M. (2000). *La sociologie de Raymond Boudon.* Paris: PUF. von Nostrand. New Hagen: Hagen University Press.

Hamlin, C. L. (2001). *Beyond relativism. Raymond Boudon, cognitive rationality and critical realism.* London: Routledge.

Nozick, R. (1981). *Philosophical explanation.* Harvard: Harvard University Press.

Perelman, C. (1982). *The realm of rhetoric.* Notre Dame: University of Notre Dame Press.

Perelman, C., & Olbrechts-Tyteca, L. (1969). *The new rhetoric: A treatise on argumentation.* Notre Dame: University of Notre Dame Press.

Perelman, C., & Olbrechts-Tyteca, L. (1986). *Trattato dell'argomentazione,* pp. xiii–xiv. Italian translation, Torino, Einaudi.

Popper, K. R. (1957). *The poverty of historicism* (p. 127). London: Routledge.

Popper, K. R. (1972). *Objective knowledge.* Oxford: Clarendon Press, New.

Popper, K. R. (1974). *Intellectual autobiography.* In P. A. Schilpp (Ed.), *The philosophy of Karl Popper.* La Salle: Open Court.

Popper, K. R. (1985). *Models, instruments and truth.* In *Selections of writings.* New York: Princeton University Press.

von Mises, L. (1949). *Human action: A treatise on economics.* Yale University.

von Mises, L. (1962). *The ultimate foundation of economic science.* Princeton: Van, Nostrand.

von Mises, L. (1976). *Epistemological problem of economics.* New York and London: New York University Press.

von Mises, L. (2011). *Socialism: An economic and sociological analysis* (1922), p. 153. Ludwig von Mises Institute: Auburn (Alabama).

Weber, M. (2013). *Economy and society (1922).* Oakland: University of California Press.

Chapter 5
Rationality, Irrationality, Realism and the Good

Paul Dumouchel

Abstract After reviewing rapidly the transformation of the concept of irrationality in its relation to our changing concept of rationality, this chapter argues that, given the normative dimension of rationality, even the most basic forms of rationality implies a conception of the good.

Irrationality and Rationality

At first sight, irrational behaviours and phenomena appear to be extremely frequent, not to say, to be the rule. Newspapers, television news reports and recorded history abound with examples of irrationality. Novelists, satirists and moralists of all ages have exploited and exposed irrational beliefs and desires. More recently, specialized research by social psychologists, economists, philosophers and others documented how supposedly sophisticated agents, like scientists or graduate students, as well as the average person often fail to meet minimum standards of rationality (Faust 1984; Tervsky and Kahneman 1986; Frank 1988; Stich 1990). As a result, there now are research projects that aim at explaining such failures of rationality, that propose different interpretations of the phenomena and distinct strategies to address the problems that arise from them, for example, (Gigerenzer and Selten 2001; Ross 2005; Kahneman 2011). The evidence now seems incontrovertible; human beings are a lot less rational than previously thought. The idea that most people are rational and act rationally most of the time is, it seems, simply false. Homo economicus is not a rational agent, or, if you prefer, behavioural economics along with cognitive science and social psychology have now showed that economic agents fail to satisfy the idealized and unrealistic requirements of perfect rationality imbedded in economic theory and shared by many approaches in social sciences.

P. Dumouchel (✉)
Graduate School of Core Ethics and Frontier Sciences,
Ritsumeikan University, Kyoto, Japan
e-mail: dumouchp@ce.ritsumei.ac.jp

© Springer International Publishing AG, part of Springer Nature 2018
G. Bronner and F. Di Iorio (eds.), *The Mystery of Rationality*,
https://doi.org/10.1007/978-3-319-94028-1_5

Yet, until recently many philosophers held the view that irrationality is a conceptual impossibility, a phenomenon that cannot exist according to their theory of rationality (Ogien 1993; Lagueux 1996, 2010)[1] and most economist held a theory of rationality which also implied that irrationality was impossible, though they did not always explicitly claim that to be case (Dumouchel 2005). Furthermore, whether irrationality is impossible or not, if phenomena are, as Vincent Descombes suggested, "that which can refute our speculations and bring us to modify our original descriptions" (1995: 9),[2] then it seems that for a long while in economics and in rational choice theory at least irrational phenomena did not exist.

Why? Consider first the methodological recommendation originally proposed in a famous essay by Milton Friedman (1953) "The Methodology of Positive Economics". According to Friedman, scientific hypotheses should not be judged on their assumptions, but on the basis of their implications, that is of their consequences and results. In the present case, this was interpreted to mean that whether economic agents actually were or *really* are rational is beside the point. What is important is, if such a hypothesis leads to empirically verifiable predictions and allows us to explain observed behaviours. In other words, economists were encouraged to overlook direct counter-examples falsifying their basic hypothesis concerning the rationality of agents, as long as that hypothesis could lead to ascertainable results.

A further example, and development of this same attitude towards empirical evidence can be found in Stigler and Becker (1977). According to this very influential article, when faced with an apparent instance of irrationality social scientists should opt for the following strategy: re-describe it in such a way that the incriminated action turns out to be rational under the new description. For example: faced with an intransitive order of preferences, revealed preference theorists usually conclude that their partitioning of outcomes was not fine enough. Thus, if someone prefers in the following order (1) an apple to pear, (2) a pear to a banana and (3) a banana to an apple, the social scientist should not jump to the "hasty conclusion" that this individual's preferences are intransitive, but adopt the more reasonable hypothesis that the outcomes over which the agent's preferences bear have not been sufficiently well defined. Thus, the outcomes or alternatives of the agent's ordered preferences should not be construed simply as apples, pears or bananas, for in that case she would be irrational, but as 'an apple against a pear', 'an apple against a banana', and so on, so that one can prefer first 'an apple to a pear', then 'a pear to a banana' and finally 'a banana to an apple' with no contradiction or irrationality whatsoever.[3] Given a minimum of ingenuity (and the absence of experimental

---

[1]Mele (1987) mentions among others Plato (Socrates), Hare (1963), Davidson (1970), Pugmire (1982), Watson (1977) as philosophers who tend to view irrationality, in particular in the forms of weakness of the will or self-deception, as a conceptual impossibility.

[2]Translation by me, Paul Dumouchel: "ce qui peut donner un démenti à nos speculations et nous amener à corriger nos premières descriptions".

[3]See Broom (1990) for an illustration of this strategy.

constraints),[4] this operation can always be carried out successfully, if not convincingly. It follows that though irrationality may exist very few instances of irrationality will ever be found. From this point of view, irrational behaviours barely constitute phenomena; they fail to refute our speculations. At best, apparent irrationality can bring us to modify our original descriptions, sometimes in an interesting way.[5]

Underlying these methodological prescriptions is a deeper issue that is clearly revealed by the attitude towards irrationality of those who adopt a weak form of the principle of rationality rather than the stronger concept of rationality, implicit in the discussion so far. According to these authors,[6] "rational action" is more or less equivalent to meaningful action or to "action" as opposed to reflex or instinctive motion. In consequence, irrational action is something of a self-contradiction, an oxymoron. Rationality in this context is not viewed as a hypothesis that could be proven false and it is more than a methodological recommendation which should be adhered to in the face of prima facie counter-examples. Rather it is something of an axiom or principle which one must postulate in order to render individual agents' behaviour meaningful.[7] Viewed in this way rationality is somewhat like Davidson's (1986) principle of charity, according to which an agent's utterances should be considered truthful and his actions meaningful, for if they were not for the most part truthful and meaningful we would not be able to make any sense of them. Rationality thus considered is a necessary precondition or presupposition to the understanding of action and that is why we cannot give it up.[8] That is why instances of irrationality must be shown to be apparent, rather than real. Irrationality may exist, but it is no phenomena; it is too obscure and incomprehensible to be able to refute any hypothesis explaining human behaviour.

More recently Ross (2005) claimed that the explanation of behaviour in either biology or in social sciences requires one to adopt an intentional-functional stance and that such explanations of behaviour will necessarily be done in terms of reasons and maximization. That is to say, behaviour makes sense to the extent that it can be described as maximizing (or minimizing) a variable in relation to ordered preferences under constraints of scarcity. In other words, his claim is that something closely akin to the standard model of economic explanation is found in any

---

[4]One of the interests of the research pioneered by Amos Tversky and David Kahneman is that it rests on an experimental design that rendered highly difficult such re-descriptions of the counter examples.

[5]An example of a creative re-description that is informative rather than purely ad hoc is Russell Hardin (1995), which argues that much of the interethnic violence that took place in ex-Yugoslavia was not to be explained by recourse to irrational fears and hatred, but was best seen as the expression of individual rationality in particular circumstances. Interestingly, Hardin also views group conflicts as instances of solutions to the collective action problem.

[6]For example, Popper (1967), Boyer (1992), Lagueux (1996), or von Mises (1944).

[7]Even if, as Popper (1967) paradoxically claims, it is patently false!

[8]For example, J. -P. Derriennic (2001) argues that if violence is not rational it cannot be understood and the political scientists must give up doing science.

explanation of behaviour, whether it is the behaviour of neurones, of animals or of enterprises. A first sight, this seems very close to the claim that the hypothesis of rationality constitutes a precondition for the explanation of action, but extended to include non-human behavior. Actually, Ross's claim represents an interesting twist in the discussion of the relations of rationality and irrationality, for according to him this requirement of intelligibility does not imply that humans are rational or act rationally in a normative sense. In fact, Ross argues, as does Glimcher (2003), that humans unlike neurones and many animals are rather imperfect maximizers, and that the use of this type of explanation is more straightforward in the natural world than in human affairs.

The extension of the theory of games to biology since the seminal work of Maynard-Smith (1982) and the subsequent development of evolutionary game theory paved the way for this transformation of the meaning of rationality. As evidence by the idea of bounded rationality as an adaptive tool box that Gigerenzer and his collaborators defend (Gigerenzer and Selten 2001; Gigerenzer et al. 2011) as well as by a recent book of de Sousa (2007) rationality now tends to be construed as one of a family of strategies, of which natural selection is the main representative, that augment the likelihood that an agent—a gene, an organism, an artificial device or a person—will reach its preferred objective. Characteristic of this transformation is the fact that certain behaviours which were previously seen as irrational, for example biases and simple heuristics, are now viewed as part of the "adaptive tool box" of bounded human rationality.

At first sight there does not seem to be that much distance between the first and the last positions in this evolution.[9] Those who argued that irrationality was impossible did not contest the existence of apparently irrational behaviour, they simply argued that these behaviours were not *really* irrational. Those who now claim that such behaviours are part of our bounded rationality rather than strictly irrational are apparently saying something that is relatively similar: that the behaviours in question are not irrational. However, the meaning of their 'not-being-irrational' is different in the two cases. For example, when Hare (1963) argues that weakness of the will is impossible, he does not claim that agents do not sometime choose a course of action which they later come to judge to have been inappropriate, but that since the agent did x he must have judged that x was the best thing to do at that time and therefore the agent did not act irrationally. He was perhaps wrong to believe that this was the best action in the circumstances, but being wrong or having a false belief is not of itself a sign of irrationality. To the opposite, advocates of the adaptive tool box claim that many biases and errors in reasoning are part of our bounded rationality, because they give sufficiently good results. That is to say, the claim is that such biases are rational, rather than simply not irrational, and these two claims are quite different.

---

[9]There is something like an evolutionary process involved here in the sense that new positions appear and tend to become dominant, though it is still possible to find authors who adhere to any one of the positions mentioned above.

In the remainder of this chapter I want to enquire into the reasons for this transformation. How did irrationality, which was first viewed as impossible and then as falsifying the claims of reason to explain human behaviour come to be included within rational behaviour? Not so long ago, irrationality was construed as either impossible or irrelevant to the explanation of human behaviour. Then, failures of rationality came to be perceived as a challenging the economic explanation of action. More recently they are being co-opted as part of, or, as examples of rationality for limited beings, while what seems to have been the core of this concept of rationality is now being applied, and is claimed to be more appropriately applied, to non-human behaviour. How did this transformation take place? What change took place in the meaning of rationality to allow this transformation?

There is of course an apparently evident answer to the questions I am asking. It is simply that the concept of rationality long cherished by economists was wildly unrealistic and bound to be falsified sooner or later by the empirical evidence amassed by psychology and behavioral economics. Conventionalist strategies notwithstanding, idealizations such as perfect knowledge or a complete ordered set of unchanging preferences came to be seen as too unrealistic to have any empirical relevance, and were viewed as obstacles to rather than means of scientific progress. Taking this evidence seriously changed rationality from an ideal property of idealized decision makers, to how real people decide in situations of uncertainty and incomplete knowledge. Rationality became naturalized. There is much truth in this deflationary answer. However, the questions I am asking are not historical but conceptual. Different notions of rationality entail different concepts of irrationality. What is irrational once reason is naturalized? How does the meaning of rationality change when some forms of irrationality are co-opted as part of reason? Can a cell be irrational?

Normativity and Realism

Reason and rationality are normative concepts. It is true that, as mentioned above, Ross (2005) claims that the criteria for the intelligibility of action that he puts forward does not entail that agents act rationally in a normative sense, but is there a non-normative sense of rationality? His criterion implies nonetheless a certain degree of normativity given that he argues that human often are bad at maximizing. The question this raises is where is normativity to be located, what place does it occupy in sciences like economics and decision theory? Rationality in these disciplines is a normative in the sense that when applied to actions, for example, it prescribes one or some actions only among the set of all possible actions to which a situation may lead, or when applied to decisions it determines one or some decisions as the optimal or best options currently available to an agent. This normative dimension of rationality entails that there are irrational actions which should be avoided.

Rationality, in economics, political science or decision theory however is not only prescriptive. Its role is not limited to determining how agents should act in given situations, it is also essential to explaining markets, voting trends, international politics, consumers' choice, in short economic and social behaviour in general. Rationality in these disciplines is both a prescriptive normative concept and a descriptive explanatory concept. It explains how agents act and it tells them how they should act. This duality, which is fundamental to normative economics, does not stem from conceptual confusion. Any theory that puts forward conditional imperatives, for example methodology, must aim at descriptive adequacy or embrace magic. This is a necessary feature of any theory which issues hypothetical imperatives. Normative theories which have no hold on reality are doomed to fail. The particularity of a theory which prescribes rational actions is that this descriptive requirement concerns not only the world in which the agents act, but also the agents for whom the prescriptions are intended. Its normative recommendations should not be beyond the means of those to whom they are addressed. Rational choices and behaviour can only be proposed to those who already are rational to a sufficient extent. It may be argued that this is precisely what happened to the conception of rationality that requires in order to be applied perfect knowledge and a set of well-ordered unchanging preferences. Its prescriptions where void because they were inaccessible to the finite beings that we are.

Rationality in prescriptive theories of action and decision appears then both as a goal to be pursued and as a condition for attaining that goal. These two dimensions of the concept cannot be separated. You cannot have one without the other. This dialectic of the descriptive and normative is clear, for example, in the discussion surrounding the adaptive toll box in (Gigerenzer and Selten 2001). In that book, over and over again the results obtained by simple heuristics and spontaneous biases were deemed to be as good or good enough when compared to those obtained by more sophisticated methods involving probabilities or calculating expected utility. Given the speed and simplicity of such heuristics, the authors argue that they are more appropriate for the task at hand than trying to satisfy the strong requirements of optimization. This argument in favour or the "adaptive toolbox" boils down to a simple cost benefit analysis of the instrument that should be used to resolve problems. This is instrumental rationality applied as a normative criterion to bounded rationality—the adaptive toolbox—conceived as an instrument for problem solving.

Where does the normative authority of instrumental rationality which is here taken for granted come from? Many authors argue that instrumental rationality is not normative, at least not in a strong sense or not in the *real* sense, because it does not contain any conception of the good, more precisely because the instrumental rationality of an act is determined independently of the "good" or the end it pursues. Instrumental rationality can be made to serve either admirable or monstrous ends. That is certainly the case, nonetheless instrumental rationality remains at least minimally normative to the extent that it recommends certain actions and discommends others. Where does its authority to do that come from?

Jean Hampton in a very interesting, but unfortunately unfinished posthumous book (1998) argues, contrary to the majority opinion, that instrumental rationality is normative and contains a definite conception of the good. That claim is part of a larger argument that aims at refuting ethical skepticism, the idea that the predicates "good" and "bad" are fundamentally strange or metaphysically "queer". Because good and bad are non-natural properties, properties that are not to be found in the world, ethics, it is argued, cannot be a real form of knowledge for there are no object (or properties) that corresponds to its basic categories. Hampton's strategy is not to reject the skeptics claim concerning the non-natural character of the good, but to argue that the normativity which is reproached to ethics is also present in science, more precisely in our conception of instrumental rationality. In particular, she argues that the axioms of expected utility theory in the von Neumann Morgenstern version impose a definite conception of the good on the preferences of agents. Since satisfying these axioms is equivalent to maximizing expected utility, it follows that acting rationally is to be guided by a conception of the good. The claim that instrumental rationality is independent from any particular conception of the good is thus false, and different conceptions of instrumental rationality, says Hampton, imply different conceptions of the good.

More generally her argument is that instrumental rationality cannot, so to speak, be instrumental all the way down, that instrumental justifications of instrumental rationality inevitably fail, because they are ultimately forced to appeal to a non-instrumental justification. My goal here is not to defend Hampton's overall project, which formulation in any case remains significantly incomplete, nor to evaluate her analysis of von Neumann Morgenstern expected utility theory as a theory of instrumental rationality. Rather I am interested in where Hampton locates the source of the normative authority of (instrumental) reason. In a conception of the good that is external to instrumental rationality and grounds its authority is the evident and straightforward answer. However, how does this conception which concerns the structure of the good, rather than its content, take shape and become visible to us? What kind of phenomena is it? These are the questions that interest me.

At the heart of her argument is the simple fact that we have many goals, numerous preferences and that we cannot act rationally, that we cannot answer questions concerning the best means to satisfy any one of these goals without taking into account at least some (perhaps all) of the other goals that we have, without in other terms imposing some kind of order on our preferences. This is a requirement that is internal to instrumental rationality, a necessary condition to be able to determine what is the best means to achieve a goal, any goal, whatever that goal may be. According to Hampton, this order which we impose upon our preferences inevitably corresponds to or constitutes a conception of the good that guides our use of instrumental reason, a conception which is not itself instrumental and that cannot be reduced to mcta-preference or second order preference.

For example, to include in the evaluation leading to a present decision one's future preference, that is a preference that I do not have now but that I consider likely that I will have in the future, cannot be reduced to having a second order preference for being "prudent" rather than "heedless".

...the person who has a second-order preference believes that the future preference is part of her good – and indeed, *that is why she has the second order preference*. Such second order preference is not like a preference for cheetos over corn chips; it is a preference that comes from reflection about how to understand and pursue her good ... the second order preference is being driven by a belief about what she *ought* to consider part of her *Source Set* of preferences if she is to act rationally... (Hampton 1998: 197)

To this the revealed preferences theorist may want to answer that how an agent's preference comes about is irrelevant as long as her set of preferences satisfies certain conditions of consistency and order she acts rationally. If an agent has second-order preference concerning her future preferences then she acts rationally by taking them into account, if she does not have such second-order preference then she also acts rationally if she does not take them into account. This response however brings us back to Hampton's analysis of expected utility theory, the constraints of order and consistency on preferences are neither first-order nor second-order preferences of agents, but axioms that impose a form on the agent's set of preferences, these axioms she argues embody a specific conception of the good. Is it possible to act instrumentally rationally without such constraints upon one's preferences?

Consider the following case introduced by de Souza (2007) to illustrate what he names categorial rationality. Andrea Yates in 2002 was condemned to life imprisonment for having drowned her five children. The defence did not contest the facts, but pleaded not guilty for reasons of insanity.

Andrea Yates had acted under the influence of voices from God, commanding her to drown her children in order to save them from Satan. This defence was rejected because she had executed her project with deliberate method, proving that she was fully aware of what she was doing. Since she had acted *rationally* in the light of the goal she had set herself, the jury inferred she could not be insane ... This basic sense of rationality is admittedly trivial, even tautological, since it follows from the fact that the agent does something "because she wants to." That is the sense in which categorial rationality, even in the cases of the most egregious irrationality, presupposes a minimal level of normative rationality. (de Souza 2007: 31–32)

Does it? Was this action in any way rational? Was it instrumentally rational, even if it was executed with method? Or should we consider that the members of the jury were wrong not only when they inferred that Andrea Yates was not insane, but when they thought she acted rationally? "Any judgement of rationality or of irrationality," adds de Souza, "depends on a given framework of relevant considerations. In the context of the most restricted framework, all action is by definition rational." (2007: 32) Hampton's argument can be understood as claiming that this extension of the framework of relevant considerations is a far from trivial operation because it is equivalent to elaborating a conception of the good.

De Souza's claim however brings us back to a point that was raised earlier in this chapter. If in the context of the most restricted framework "all action is by definition rational," how is irrationality possible? And if irrationality is impossible, in what sense can we understand instrumental rationality to be even minimally normative? The same question can be asked in a different way: is it enough that an action be a purposeful in order for it to also be instrumentally rational?

Consider the following dream: "you are in a car driving with a friend you come to a corner and turn right to go downtown, then you are on a beach with someone else having a heated discussion, a large building comes in sight and now you are walking up a hill with other soldiers, a strange aircraft passes overhead and there is an explosion in the distance you do not hear it but you see the smoke, as you turn a bend in the road you can see where the explosion was, far away, and from it a flow of incandescent material is coming down the hill towards you, you run to escape it and wake up just before it reaches you."

There is purposeful action here. You do somethings—you turn left and run away from danger—for a reason, or because "you want to." Other events however simply seem to happen to you, you are driving, you are on the beach, you are walking with other soldiers. Strangely enough the transformation, or transportation, without transition from one situation to another do not seem to bother you. If awake one day as you got up from your chair you found yourself in a completely different location from where you were when sitting, for example in a boat instead of your office, this would probably not go unnoticed to say the least. However, there are other events and actions which are of concern to you: going downtown, the heated discussion, the flow of incandescent material that you want to avoid. Unlike being transported immediately and without reason from one place to another completely different and unrelated location, this last transformation of the situation leads you to wake up.

Clearly something is missing here, something is absent from the dream and that may be called either the World or realism. Realism in this sense is the stance that my representations are of what does not concern me and of what has no concern for me: the world. Realism so understood, as Bimbenet (2015) argues, is not a philosophical position or doctrine, but a cognitive stance characteristics of the type of animals that we are. When I was an undergraduate student the professor who introduced skepticism began by saying that skeptics are not people who fail to go around tables when they see them because they are not sure if they are there or not. Realism in the sense in which it is used here is what makes skeptics, like all of us, walk around the tables they see. That is, the assurance that objects exist in themselves, independently of us and that the relations between them or the rules that govern their behaviour are just as real and independent of us as the objects are. The World is the totality of all that exists in that sense and in any other sense of 'exists'.[10] That is why we can never represent it as such. It is only given to us through the stance of realism, its expectations and presuppositions. The world is an imagined object.

Realism as a stance constructs for us the World as a totality that is different from the worlds that we (or that biologists) attribute to animals in that this totality is not centered around the animal that we are, is not for us in the way in which the world of an ant or of a chimpanzee is for the immediate needs of the ant or the chimpanzee. The World is in itself. Realism in that sense is what is missing from dreams

---

[10]Because we may be wrong and often are wrong about what is and the relations between what is, it is clear that neither realism nor the world are equivalent to what is real.

like the one described above.[11] We could say that in such dreams, to paraphrase Kant, "the *I dream* constitutes the transcendental unity of all its representations." The only thing that provides the unity to all that happens is that all that happens concerns me, the dreamer. All that happens is related to me and what is not, for example that I go without transition from driving a car to walking on a beach, does not deserve attention or require explanation, and in that sense does not even happen. All representations in the dream gain their unity from being mine and that is the only thing that all the events of the dream have in common and need to have in common.

The dreamer can act purposefully, but she cannot act instrumentally, if that implies rationality, because she has no world. Not only the circumstance may change radically before the action is carried out, but the means may turn out to be inadequate for the task at hand because as it is being done the immediate conditions of the task evolve frustrating the agent's endeavours. An action can be purposeful without being rational, because rationality implies realism, it adverts to the world, but purposefulness does not need the World, it only needs one object, its target. That is why, like classical empiricism, purposefulness is essentially subjective, condemning the agent to the prison of his representations. An organism that follows a unique purpose has a goal; it can even be described as acting for a reason, but it is not rational.

Realism and the Good

As evidenced by many recent works, for example (Gigerenzer and Selten 2001; Hammerstein 2003; Glimcher 2003; Ross 2005; de Souza 2007) there is a tendency in studies on rationality to insist on the proximity between rationality and natural selection. Both give rise to teleological action, and, as mentioned earlier, both belong to a family of strategies that are means which promote an organism's chance of succeeding in its enterprises or if you prefer of reaching its preferred goal. An organism that pursues a single goal blindly, like the bacteria that follows the earth magnetic field and thus avoid surface water that is rich in oxygen, does not need to have a world. The totality of all that exists for such bacteria is the magnetic field that guides it and the food it encounters. However, if it does not need *to have a world*, if it can survive and succeed with a very limited domain of perception or representations, it needs *the world*. That is to say, it needs a stable environment where the conditions of its successful action are satisfied. That is why when the environment in which its adapted behaviour changes the behaviour fails to reach its "intended" goal. Because natural selection has, so to speak, externalized most of the conditions of the organism's successful action, such an organism cannot have a

---

[11]I do not claim that all dreams like the one described above, lack a world in that sense, though I suspect that all do. It is enough for my argument that some dreams are like that.

plurality of goals. More an organism has goals and more it needs to have a rich and complex world, that is to say, more it needs to represent adequately the elements or properties of the world that are relevant for achieving its goals.

Realism however is different; while the world of most animals centers around the animal, and in that sense is radically subjective, the realist stance asserts the existence of the World, a world of which I am not the center and which is objective rather than the mere sum of my representations. Once again this is not a philosophical or metaphysical claim. The contention is not that we humans, unlike other animals, manage to reach reality as it *really is*, and to escape the limits imposed upon us by our evolutionary history. Rather it is that we spontaneously adopt a cognitive stance that assumes the existence of a self-sustaining independent reality in which we are included as an element among others, a reality that is ordered (organized or structured) in a way that is independent of our goals and preferences. This is quite different from the world that we attribute to animals, whether it be the *umwelt* of Uexküll (1956) or the domains of cognition attributed to them by cognitive ethologists (Allen and Bekoff 1999; Lurz 2009). These animal worlds are centered upon the animal, they are cut out of the reality in which we live by the animal's goals and survival. It may be of course that biologists are simply wrong about this, and that animals are realists just like us. It may be... but there is a lot of evidence concerning animal behaviour, even that of our closest cousins, that suggests that this is not the case. Chimpanzees that are taught sign language and succeed in mastering it have very little to say and, in particular, they never say anything that does not concern them directly. They ask to go out, for food or for something else they want, but they do not draw the attention of their addressee to the presence of an unfamiliar object, ask for information or explanation, or refer to the past. Further as Vauclair (1995: 125) reports, an important difference in the communicative exchange between mother and infant among humans and similar exchange among other primates is the inclusion of objects in human communicative exchange. Like our dreamer, animals seem to be primarily attentive to and interested in what concerns them directly or what is relevant to their needs. In relation to the objective world which we inhabit, they live as in a dream.

Realism, in the sense in which we humans are spontaneous realists is not simply equivalent to having a plurality of goals. It is to have a plurality of goals that are not structured and made coherent by the world in which we live because the world in which we realists live is not centered around us. Reason, I agree with Hampton, is a means to bring coherence and order among our various goals and preferences. The first step in order to do that is change from a spontaneous realist to a determinate or willful realist. It is, to put it otherwise, to take into account more and more of the structure of the world that is not structured around us. It is, for example, to take my future preferences into consideration when deciding about my present options, for that I will grow old and that as I do my preference will change is as real as the chair on which I am now sitting. Taking reality into account, including in our process of decision the other goals and preferences that we have and how our various options will react upon them is the basic condition of what we call instrumental rationality, the basic condition for us to achieve as well as possible as many of our goals as we can. How far that inquiry should be pursued is an open question. It may receive depending on the person and circumstances different answer which may all be just as reasonable.

However, if Hampton is right that the order and constraints which we impose upon our set of preferences embodies a conception of the good, it follows that at the most basic level, what constitutes the normative element present in instrumental rationality is being attentive to world and receptive of reality. That is also a fundamental component of the good, in the sense that it is good. The good so understood is neither a property of the world, nor something that we project upon it. Rather, it is more like an affordance according to Gibson, an opportunity which we may seize. That is the good of which Andrea Yates could not take advantage.

References

Allen, C., & Bekoff, M. (1999). *Species of mind the philosophy and biology of cognitive ethology.* Cambridge, MA: MIT Press.

Bimbenet, E. (2015). *L'invention du réalisme.* Paris: Les éditions du Cerf.

Boyer, A. (1992). *L'Explication en histoire.* Presses Universitaires de Lille.

Broom, J. (1990). Should a rational agent maximize expected utility. In K. S. Cook & M. Levi (Eds.), *The limits of rationality* (pp. 132–145). Chicago: University of Chicago Press.

Davidson, D. (1970). How is weakness of the will possible? In *Essays on actions and events* (pp. 21–42). Oxford: Oxford University Press, 1980.

Davidson, D. (1986). *Inquiries into truth and interpretation.* Oxford: Oxford University Press.

Derriennic, J. P. (2001). *Les Guerres civiles.* Paris: Les presses de la fondation nationales des sciences politiques.

Descombes, V. (1995). *La denrée mentale.* Paris: Minuit.

de Souza, R. (2007). *Why think? Evolution and the rational mind.* Oxford: Oxford University Press.

Dumouchel, P. (2005). Rational Deception. In C. Gerschlager (Ed.), *Deception in markets* (pp. 51–73). New York: Palgrave Macmillan.

Faust, D. (1984). *The limits of scientific reasoning.* Minneapolis: Universit of Minnesota Press.

Frank, R. H. (1988). *Passions within reason.* New York: W.W. Northon.

Friedman, M. (1953). The methodology of positive economics. In *essays in positive economics* (pp. 3–43). Chicago: University of Chicago Press.

Gigerenzer, G., & Selten, R. (Eds.). (2001). *Bounded rationality the adaptive toolbox.* Cambridge, MA: MIT Press.

Gigerenzer, G., Hertwig, R., & Pachur, T. (Eds.). (2011). *Heuristics: The foundation of adaptive behavior.* Oxford: Oxford University Press.

Glimcher, P. (2003). *Decision, uncertainty and the brain.* Cambridge, MA: MIT Press.

Hammerstein, P. (Ed.). (2003). *Genetic and cultural evolution of cooperation.* Cambridge, MA: MIT Press.

Hampton, J. (1998). *The authority of reason.* Cambridge: Cambridge University Press.

Hardin, R. (1995). *One for all the logic of group conflict.* Princeton: Princeton University Press.

Hare, R. (1963). *Reason and freedom.* Oxford: Oxford University Press.

Kahneman, D. (2011). *Thinking fast and slow.* London: Penguin Books.

Lagueux, M. (1996). How could anyone be irrational. In M. Marion & R. S. Cohen (Eds.), *Québec studies in the philosophy of science II* (pp. 177–192). Dordecht: Kluwer Academic Press.

Lagueux, M. (2010). *Rationality and explanation in economics.* London: Routledge.

Lurz, R. W. (2009). *The philosophy of animal minds*. Cambridge: Cambridge University Press.

Maynard-Smith, J. (1982). *Evolution and the theory of games*. Cambridge: Cambridge University Press.

Mele, A. (1987). *Irrationality*. Oxford: Oxford University Press.

Ogien, R. (1993). *La faiblesse de la volonté*. Paris: Presses Universitaires de France.

Popper, K. (1967). «La rationalité et le statut du principe de rationalité». In E. Classen (Ed.) *Les fondements philosophiques des doctrines économiques* (pp. 142–150), Paris: Payot.

Pugmire, D. (1982). Motivated irrationality. *Proceedings of the Aristotelian Society, 56,* 179–196.

Ross, D. (2005). *Economic theory and cognitive science*. Cambridge, MA: MIT Press.

Stich, S. P. (1990). *The fragmentation of reason*. Cambridge, MA: MIT Press.

Stigler, G. J., & Becker, G. (1977). De Gustibus Non Est Disputandum. *The American Economic Review, 67*(2), 76–90.

Tversky, A., & Kahneman, D. (1986). Rational choice and the framing of decisions. *Journal of Business, 54*(4), 251–278.

Uexküll, J. V. (1956). *Mondes animaux et monde humain*. Paris: Denoël.

Vauclair, J. (1995). *L'intelligence animale*. Paris: Seuil.

von Mises, L. (1944). The treatment of 'irrationality' in social sciences. *Philosophy and Phenomenological Research, 4*(3), 544–559.

Watson, G. (1977). Skepticism about weakness of the will. *Philosophical Review, 86,* 316–339.

Chapter 6
First Generation Behavioral Economists on Rationality, and Its Limits

Roger Frantz

Abstract The foundations of neoclassical theory came under sustained attack beginning in the 1940s. The foundations are rationality, maximization, and (allocative) efficiency. The sustained attack came from "first generation behavioral economists."

Introduction

The foundations of neoclassical theory came under sustained attack beginning in the 1940s. The foundations are rationality, maximization, and (allocative) efficiency. The sustained attack came from "first generation behavioral economists." Members of this group may vary according to the person writing about the first generation. For purposes of this paper the members are George Katona, Harvey Leibenstein (Frantz 2007, 2017), Herbert Simon (Frantz 2016a), and Frederick Hayek (Frantz 2013). Their philosophies of rationality, the place of the concept of rationality in their writings, and their view of the importance but also the limits of rationality are the topics of this chapter. Hayek is rarely if ever discussed in the context of behavioral economics. However, he was writing on topics which were similar to the writings of Katona, Leibenstein, and Simon.

The Foundations of Neoclassical Theory

We perceive that neoclassical theory is based on three concepts: complete rationality maximizing behavior, and (allocative) efficiency. People are rational, which means that they maximize (something), and the outcome is efficiency.

R. Frantz (✉)
San Diego State University, San Diego, CA, USA
e-mail: rfrantz@mail.sdsu.edu

© Springer International Publishing AG, part of Springer Nature 2018
G. Bronner and F. Di Iorio (eds.), *The Mystery of Rationality*,
https://doi.org/10.1007/978-3-319-94028-1_6

Rationality is the way people think, maximization is the way they (rationally) pursue what interests them, and efficiency is the outcome. Maximization is the intervening variable. The First Generation questioned the tripartite foundation of neoclassical economics and caused future economists to question it as well. We will discuss each of these in turn.

Rationality

In the *Manual of Political Economy*, Pareto reveals just how limited are the human actions which constitute political economy, and the role of rational or logical behavior.

"We will study the many logical, repeated actions which men perform to procure the things which satisfy their tastes… we will consider only repeated actions to be the basis for claiming that there is a logical connection uniting such actions. A man who buys a certain food for the first time may buy more of it than is necessary to satisfy his tastes… But in a second purchase he will correct his error, in part at least, and thus, little by little, will end up procuring exactly what he needs. We will examine this action at the time when he has reached this state. Similarly, if at first he makes a mistake in his reasoning about what he desires, he will rectify it in repeating the reasoning and will end up by making it completely logical… The study of these actions makes up the subject of political economy" (Pareto 1971, p. 103).

Pareto defined economics as consisting of only logical or rational behavior. Are people rational? In Pareto's theory people are rational by definition. Selective or bounded rationality are, by definition, excluded. What is the role of rationality in Pareto's theory? It "simplifies the problem enormously" (Pareto 1971, p. 103). Simplification makes our work easier. Quotes about the importance of rationality in economic theory by two first generation behavioral economists: Katona and Simon.

"The principle of rationality occupies a central position in traditional economic theory. It states that the rational man aims at a specific end—the maximization of utilities—by employing well-defined means, namely, the weighing of available alternatives" (Katona 1975, p. 217).

"Macroeconomists assume that the economic actor is rational, and hence he makes strong predictions about human behavior without performing the hard work of observing people… Thus, the classical economic theory of markets with perfect competition and rational agents is deductive theory that requires almost no contact with empirical data once its assumptions are accepted" (Simon 1959, p. 254).

"Economic man has a complete and consistent system of preferences that allows him always to choose among the alternatives open to him; he is always completely aware of what these alternatives are; there are no limits on the complexity of the computations he can perform in order to determine which alternatives are best; probability calculations are neither frightening nor mysterious to him" (Simon 1997, p. 87).

Lionel Robbins says this about rationality's role in economic theory.

"And… in the last analysis Economics does depend, if not for its existence, at least for its significance, on an ultimate valuation—the affirmation that rationality and ability to choose with knowledge is desirable. If irrationality, if the surrender to the blind force of external stimuli and uncoordinated impulse at every moment is a good to be preferred above all others, then it is true the *raison d'etre* of Economics disappears" (Robbins 1932, p. 141).

Maximization

Rationality is related to another foundation of neoclassical theory: maximization: rational behavior results in maximization (of something). Herbert Simon, again: "Just as the central assumption in the theory of consumption is that the consumer strives to maximize his utility, so the crucial assumption in the theory of the firm is that the entrepreneur strives to maximize his residual share—his profit" (Simon 1959, p. 262).

In his Nobel Lecture, "Maximum Principles in Analytical Economics," Paul Samuelson, reminds us that the principle of maximization is not new to economics.

"Alfred Marshall's Principles of Economics, the dominating treatise in the forty years after 1890, dealt much with optimal output at the point of maximum net profit. And long before Marshall, A. A. Cournot's 1838 classic, Researches into the Mathematical Principles of the Theory of Wealth, put the differential calculus to work in the study of maximum-profit output. Concern for minimization of cost goes back… at least… to the marginal productivity notions of von Thünen" (Samuelson 1970, p. 2).

Efficiency

Efficiency, market allocative efficiency, another of the foundations, is the result of maximizing/rational behavior. Efficiency, like maximization, is achieved when there exists, as Arrow stated it, "marginal equalities"—$MUx/Px = MUy/Py$, $MPl/Pl = MPk/Pk$, $MR = MC$. Efficiency is so central that Robert Mundell, the 1999 winner of the Nobel Prize, lamented about studies showing that (market allocative) inefficiency is an extremely small percentage of GDP, perhaps 0.001 or 0.0001 of GDP. Mundell summarizes his feelings about this in these words:

> …there have appeared in recent years studies purporting to demonstrate that…gains from trade and the welfare gains from tariff reduction are almost negligible. Unless there is thorough theoretical re-examination of the validity of the tools on which these studies are founded…someone inevitably will draw the conclusion that economics has ceased to be important!". (Mundell 1962, p. 622)

For the entire economy allocative inefficiency has been estimated at between 0.001 and 0.0001 of GDP; 1/10 of 1% and 1/100 of 1% of GDP. For a \$16 trillion GDP this is equal to between \$16,000,000,000 and \$1,600,000,000. By way of comparison, each year Americans spend $7,000,000,000 on potato chips. Allocative inefficiency is, to use a line from the *Godfather*, "small potatoes." It is, for all intents and purposes, insignificant. Hence Mundell's lamentation.

Rationality—Maximization—Efficiency

The three legs of neoclassical theory. People are rational, which means that they maximize (something), and the outcome is efficiency. Rationality is the way people think, maximization is what they (rationally) pursue, and efficiency is the outcome. Maximization is the intervening variable. Individuals are rational, we either equate at the margin, or we don't. We either use the procedures which allow us to maximize, or we don't. Individuals maximize utility, or we don't. But efficiency is not directly about individual behavior. *Markets* are allocatively efficient or they aren't. Either $P = MC$, or it isn't. *Firms* are either X-efficient or they are not. Firms are either on their cost and output frontier or they are not. Individuals are not X-efficient. Individuals contribute to either X-efficiency or they do not. But the level of X-efficiency or allocative efficiency is the effect of the collective behavior. The sum total of individual behavior either leads to $P = MC$, or it does not. It either leads to the firm minimizing its costs or it does not. Individuals are rational or not. Firms and markets are either efficient or not. The First Generation questioned the tripartite foundation of neoclassical economics and caused future economists to question it as well.

Rationality

George Katona

George Katona was interested in the process or procedures people use in decision making. Is the *process* a rational one? By contrast, economic theory emphasized the rationality of the outcome, e.g., did the decision maker equate at the margin. Herbert Simon called the process and outcome foci procedural and substantive rationality, respectively.

Katona described rational behavior as "purposive and understandable behavior" (Katona 1963, p. 49), behavior based on an intelligent and conscientious weighing of available alternatives. Rational behavior leads to "genuine decision making- deciding according to the requirements of the situation" (ibid., pp. 49–50). In his 1940 book *Organizing and Memorizing* Katona discusses "meaningful learning" which has similarities with genuine decision making. Meaningful learning yields an understanding of *procedures*, "insight into a situation, and "apprehension of relations" (Katona 1940, p. 5). Senseless learning is the result of memorization. Katona's research led him to conclude that forgetting occurs faster with senseless learning.

How does learning take place? In Chap. 8, "Memory Traces" he begins the chapter by saying, "In this attempt to understand the process of learning, I propose to go a step further and to resort for a short while to what is often called speculation…What takes place between the learning period and the recall?" (ibid., p. 193). When a person sees a cat for the first time it creates a pattern in the brain; the creation of the pattern is the process of memory. When the experience ends a trace of the experience remains in the brain. This trace of the memory is a *memory trace*. The next time we see a cat the memory trace is essential in our having a conscious experience of a cat. Katona says that a memory trace is a "carrier of the connection between events…" (ibid., p. 194). Memory traces "are not copies of the perceptions, and perceptions are not "summative copies of the external world" (ibid., p. 197). In other words, we do not perceive the external world as it is in reality, and what remains of the experience is not exactly identical with the true experience. This is an important point because it is similar to Hayek's thesis in *The Sensory Order*, that we do not perceive the real world as it exists.

According to Katona *homo economicus* "lists all conceivable courses of action and their consequences …ranks the consequences and chooses the best; finally he sticks to his choices in a consistent manner" (Katona 1960, p. 138). One alternative to rational behavior is *not* irrational behavior, it is habitual or routinized behavior. Human behavior is guided by habit or repetition, so much so that Katona says it is "The basic principle of one form of behavior" (ibid., p. 139). Habits result in our behavior being relatively inflexible, and hence does not result in a decision. The product of habitual behavior is an automatic response without regard to motives or needs. Katona describes it as "emotional, haphazard, nonunderstandable" behavior (op. cit.), "impulsive, whimsical… even ununderstandable behavior" (Katona 1975, p. 217). It results from behavior based on old routines, rules of thumb, group behavior and leaders without considering their relevance. It is mechanistic and 'blindly' repetitive behavior. This leads to non-genuine, non-rational, decision making. Non-genuine decision making is concomitant with non-maximization. But, "Maximization constitutes an instance of a one-motive theory, to be contrasted with the… multiplicity of motivational patterns" (ibid., p. 225). By contrast, non-habitual behavior is

> problem-solving behavior…characterized by the arousal of a problem or question, by deliberation or thinking which involves reorganization in a specific direction, by understanding the requirements of the situation, by weighing alternatives and taking their consequences into consideration, and finally, by choosing among alternative courses of action. (Katona 1960, p. 139)

Although habitual behavior is common and leads to inflexible behavior, breaking behavior habits happens when, for example, "…sometimes a bell rings; we stop, look, and listen before going along in the usual way. Habitual forms of behavior may be abandoned when we perceive that a new situation prevails and calls for different reactions" (Katona 1960, p. 140). The bell rings means that pressure has increased to transcend the habits. Deviating from habitual behavior is "one of the major features of genuine decision making" (Katona 1975, p. 221). Hence, "Only in those relatively few instances when the decision really matters do strong motivational forces arise that lead businessmen and consumers to proceed in a circumspect manner. But even in those cases the alternatives considered are usually restricted according to their attitudes and expectations, and a satisfying course of action is chosen rather than one that will optimize the outcome" (Katona 1980, p. 13–14).

But this bell ringing does not equate to rational behavior. Simply, the bell goes off and we begin to engage in *problem solving* behavior. In order of the degree of rationality human behavior can be rational, problem solving or habitual. Thus, problem solving behavior is not rational behavior. Although Katona does not this concept, problem solving is a form of bounded rationality. Katona describes problem solving in these terms.

> Rather than being a trial-and-error process in which all possible courses are listed and weighed against one another, the essential feature of problem solving, the reorganization or restructuring of the situation, is a highly selective process. By utilizing certain (not all clues from the environment, an item of information is seen in a different context and therefore in a new light; this leads to a new way of ordering of alternatives and to changed behavior. Problem solving is a flexible process rather than a rigid one. It represents adaptation by the organism to changing conditions; it represents intelligent learning, often rapid learning, rather than consistently sticking to choices once made (Katona 1960, p. 140).

Katona's own on consumer behavior concluded that about 25% of major purchases "lacked practically all features of careful deliberation. These include not considering alternatives or consequences of decisions, or not seeking information relevant to the decision. On the other hand, close to one-half of the purchases contained several features" (Katona 1975, p. 20). What are we to conclude? That individuals are (mostly) "selectively rational." That is, rationality is a continuous variable being able to take on a value between 1.0 (or perfectly rational) and 0.0 (or perfectly non-rational or perfectly irrational). Why selective rationality? Because we have *limited information* of relevant variables, live in an *uncertain environment*, and are *motivated by many things, not only maximization.* In what could also have been said by Simon or Leibenstein, Katona says of rational expectations theory that it makes "observation and measurement of expectations unnecessary by assuming that expectations are formed rationally" (Katona 1980, p. 14).

Herbert Simon

Katona is not the only one who didn't put much faith in rational expectations. Simon said that "To the best of my knowledge, all of these equations have been conceived in the shelter of armchairs; none of them are based on direct empirical evidence about the processes that economic actors actually use to form their expectations about future events" (Simon 1982a, p. 310).

Simon also did not believe people are perfectly or unboundedly rational. He distinguished unbounded rationality from bounded rationality. He also distinguished substantive from procedural rationality. In "A Behavioral Model of Rational Choice (Simon 1955, p. 99) Simon lists the characteristics of an unboundedly rational person with these words:

Traditional economic theory postulates an 'economic man,' who, in the course of being 'economic' is also 'rational.' This man is assumed to have knowledge of the relevant aspects of his environment which, if not absolutely complete, is at least impressively clear and voluminous. He is assumed also to have a well-organized and stable system of preferences, and a skill in computation that enables him to calculate, for the alternative courses of action that are available to him, which of these will permit him to reach the highest attainable point of his preference scale.

Simon rejected this in favor of bounded rationality. Bounded rationality occurs because of our *cognitive limits*. The limits of our ability to process information makes our rationality bounded, or limited. Of the entire amount of new information generated by our entire environment, maybe one percent reach our conscious mind. Therefore, a conscious consideration of costs and benefits is simply out of the question. Our degree of rationality is bounded by the relative *complexity* of the world, given our cognitive abilities. Our ability to solve problems is *as if* we "search through a vast maze of possibilities, a maze that describes the environment" (Simon 1996, p. 66). Along with his ejection of unbounded rationality, Simon also rejected its concomitant, maximization, or optimization. As a description of human behavior Simon preferred the term *satisfying*. In *Administrative Behavior* Simon says that

> Whereas economic man supposedly maximizes –selects the best alternative from among all those available to him – his cousin, the administrator, satisfies- looks for a course of action that is satisfactory or 'good enough'… Economic man purports to deal with the 'real world' in all its complexity. The administrator recognizes that the perceived world is a drastically simplify model of the buzzing, blooming confusion that constitutes the real world. (Simon 1997, p. 119)

Satisfying means having an aspiration level which is satisfactory, and choosing the first alternative which satisfies the aspiration. This two step process is also known as the *satisfying heuristic*.

Simon also distinguished substantive from procedural rationality, favoring the latter. Substantive rationality is a description of the outcome of (rational) decision making. Substantive rationality means equating at the margin. *Homo economic us* is a substantively rational maximizer of subjective expected utility. But, Simon says, "…there is a complete lack of evidence that, in actual human choice situations of any complexity, these computations can be, or are in fact, performed" (Simon 1982b, p. 244).

The second type of rationality is procedural rationality. Procedural rationality is about the *process* of choice or the procedures used to make a decision. An individual is thus procedurally rational when, *given* their cognitive limits, the deliberations leading to a decision are *appropriate*. Procedural rationality is not rationality judged by some absolute standard of the ability to think logically. It is judged given your ability to think logically, so long as your decisions are appropriate, given your limited ability to think logically. Impulsive decision making would, therefore, in the absence of an extraordinary amount of luck, be procedurally irrational. (Simon 1982c, p. 426). Rather than maximizing by equating at the margin we tend to use *pattern recognitions* which remain in our *subconscious,* and attempt for satisfying rather than maximizing results.

Pattern recognition and the subconscious were well illustrated by Simon in his discussion of chess. There are about 30 legal moves in a chess game. Each move and its response creates an average of about 1000 contingencies. In a 40 move chess game there are about 10^{120} contingencies. Chess masters are believed to look at no more than 100 contingencies. beginning with an inordinately large number of possibilities, chess masters, and humans in general search for outcomes whose utility values are at least satisfactory. Once found, the search stops. In other words, chess masters are satisfiers, and their rationality is bounded by their limited cognitive capacity relative to their environment.

Chess grandmasters take so little time to decide on a move that Simon says that it is not possible for their moves to be the product of "careful analysis" (Simon 1983, p. 133). A grandmaster takes 5 or 10 s before making a strong move, which 80–90% of the time proves to be correct and one that is "objectively best in the position" (Simon 1983, p. 25). Their skill barely diminishes when they play 50 opponents at once rather than one opponent. How do they do it? When grandmasters are asked how they play they respond with the words intuition and professional judgment. The process is subconscious pattern recognition based on experiences stored in memory and retrieved when needed. While short term memory can store only a relatively small amount of information, long term memory is, metaphorically speaking, a large encyclopedia with an elaborate index in which information is cross-referenced. Cross-referencing means that information is associative with one piece of information linked or associated with other associated thoughts. Cross-referencing and chunking makes subconscious pattern recognition or intuition easier. Human experts at chess or in any field of activity experts are expert (in part) because of their ability for subconscious pattern recognition.

Cognitive limits relative to a complex and uncertain environment are not the only reasons why rationality is difficult to achieve. For example, if there are two people making decisions who are competing against each other, then person A's rational strategy depends on person B's reaction to person A's decision. This is because person B's reaction affects the future and hence the consequences of person A's decision. Person A is rational depending on his or her ability to read B's mind. Similarly, B is rational depending on his or her ability to read A's mind. Rationality is about mind reading as much as it is about the proper use of reason. Of course, an individual would have to know how their behavior affects others which affect themselves "through unlimited stretches of time," and "unlimited reaches of space" (Simon 1997, p. 69). "Inconceivable" is Simon's description of anyone's ability to be able to do these things. Rationality is bounded by our cognitive abilities relative to the environment. Another obstacle to be overcome is solving a really large number of equations simultaneously as if the system is a "*closed system.*" A closed system has a limited number of variables because the scientist can limit them in their *laboratory* experiments. Simon says that "Rational choice will be feasible to the extent that the limited set of factors upon which decision is based corresponds, in nature, to a closed system of variables—that is, to the extent that significant indirect effects are absent" (Simon 1997, p. 83; emphasis added).

Are we ever (fully, near or quasi) rational? Simon says in a statement which could have been made by Katona, "Only in the cases of extremely important decisions is it possible to bring to bear sufficient resources to unravel a very involved chain of effects" (Simon 1997, p. 83). To put it in terms Leibenstein could say, pressure enhances rationality.

Harvey Leibenstein

Leibenstein's definition of rationality is procedural. Individuals are rational if their behavior includes the necessary procedures for the firm to be on their production and cost frontiers. He says of the neoclassical approach to rationality that it "…may be characterized by the phrase *complete constraint concerned 'calculatedness' in the pursuit of precise objectives*" (Leibenstein 1976, p. 72). Constraint concern is the willingness to be aware of or concerned about the constraints that are faced and working through the constraints in order to make a cost minimizing or output maximizing decision. In order to do this you need to be able to make a realistic assessment of the environment, make an independent judgment rather than succumb to peer pressure when you know that your judgment is superior to that of the others, be aware of relatively small changes in the environment, don't put off for tomorrow what you don't want to do today, and learn from experience. These are the processes which constitute rational behavior. According to Lebenstein these procedures are followed if the individual is 'guided' by their superego, that part of our personality which is willing to calculate, work through problems regardless of how much time and effort is required in order to do well. The alternative is the id, our 'inner surfer,' our 'inner dude.' Leibenstein says that

> an individual effects a compromise between his desire to do as he pleases and internalized standards of behavior acquired through background and environment. Thus we assume that individuals are influenced by others and that their psychology requires them to strike a balance between conflicting desires (ibid. p. 71).

The influence of others within an organization are the "almost *invisible* bonds… between individuals" (Leibenstein 1987, p. 4), the "visible and invisible *connections*" (op. cit.) which have a large effect on the firm's performance. Some of the influence of others works on the subconscious.

Because both the superego and id are part of our personality, because we are affected by others, we tend towards *selective* rationality—selectively X-efficient—rather than perfect rationality all the time, every day, 24/7. Without perfect rationality firms are neither on their cost frontier nor their production frontier. Leibenstein's x-efficiency theory thus refutes the notion that a firm's output and cost levels are determined "mechanically" by the quantity of inputs and technology, and that the firm's efficiency is dictated as if by a set of blueprints created by an engineer. Leibenstein's x-efficiency theory maintains that there is something beyond a set of blueprints which determines output and cost; Leibenstein calls this the x factor. It is the x factor which led him to name his theory x-efficiency (Leibenstein 1966). Economics is thus behavioral, not mechanical, it is more about biology, anthropology, psychology, than physics. Numerous studies from many industries in all parts of the world have shown that on average firms are considerably below their production frontier or above their cost frontier. In other words, X-inefficiently x-ists (Frantz 2016b).

Rationality is a continuous variable with an "economic person" exhibiting complete constraint concern at one extreme, and our "inner surfer dude" The economic person is thus a limiting case: a characteristic of the decision-making procedures used by some people at some times but not necessarily characteristic of all people at all times. X-inefficient behavior most likely occurs when an id-dominating personality (internal environment) is combined with a low pressure market structure (external environment), and a set of group norms which discourages or at least does not encourage high work effort. X-efficient behavior most likely occurs when an super-ego dominating personality (internal environment) is combined with a high pressure market structure (external environment), and a set of group norms which encourages or at least does not discourage high work effort. Leibenstein used the metaphor of a pair of scissors to explain human behavior. One blade is the individual's inner environment, his personality. The other blade is the external environment, the degree of market competition creating pressure.

Leibenstein's view of rationality being procedural is consistent with Simon. But Leibenstein does not adhere to the satisfying view of behavior. X-inefficiency stems from sloppiness, and laziness, not satisfying behavior. Satisfying is similar to maximizing in that both are goal oriented, directed behavior. A person who satisfies does so because they do not have the degree of knowledge necessary to maximize. However, given their limited knowledge they do the best they can. An x-inefficient person may or may not have enough knowledge to maximize. However, regardless of the degree of knowledge they have, they don't use it as well as they can. One reason for not using knowledge as well as they can is that "a good deal of our knowledge is *vague (tacit)*. A man may have nothing more than a sense of its existence, and yet this may be the critical element. Given a sufficient inducement, he can search out its nature in detail and get it to a stage where he can use it. People normally operate within the bounds of a great deal of intellectual slack. Unlike underutilized capital, this is an element that is very difficult to observe" (Leibenstein 1976, p. 41; emphasis added).

Another way of expressing this is that economics deals with *complexity*, with a large number of observations and variables and relationships among the variables. Economic events are complex because they are affected by economic and non-economic variables, and these numerous non-economic variables "cannot be accounted for on the basis of existing knowledge" (ibid. pp. 14–15). There are simply *too many possible interactions of (known and unknown) factors influencing economic behavior and the economic system.* Part of the reason for the complexity is that the "social and economic phenomena are usually not completely separable but are organically intertwined as part of entities larger than the entity being studied" (ibid. p. 19). Again, "*the system the economist is interested in is part of a larger system of relations that is unknown in its totality*" (ibid. p. 16). By contrast, "This is precisely the sort of thing the *laboratory scientist* is able to get away from by creating in the laboratory an artificial environment whose state he is able to control" (ibid. p. 15). While a scientist in a laboratory can create a more simple environment, economists have no such luxury: their environment is too complex. Economists and other social sciences can't predict specific events at specific times, but we can explain "general trends" (ibid. p. 21). Hayek calls it the "explanation of the principle." The existence of a time dimension also leads Leibenstein to maintain that prediction is difficult, even impossible. Leibenstein sounds Hayekian when he says that a system that will predict what will actually happen – is, in principal, impossible. Even if we knew all the necessary initial data, as the system unfolds the environmental parameters would change; they would; g influence some of the variables within the system and the results would not be in accordance with what we would have predicted at the outset (ibid. p. 15).

Frederick Hayek

Herbert Simon says of Frederick Hayek that he was essential in the development of the concept of bounded rationality. In *The Sciences of the Artificial* Simon stated that "No one has characterized market mechanisms better than Friedrich von Hayek [...] His defense did not rest primarily upon the supposed optimum attained by them but rather upon the limits of the inner environment—the computational limits of human beings" (Simon 1996 p. 34).

Hayek opposed the idea that rationality means making logical deduction from explicit premises. Hayek preferred defining rationality as the ability to "recognize truth... when [one meets] it..." (Hayek 1967, p. 84). A second definition of rationality is the ability to be understood by others. A rational person's behavior follows a pattern which can be understood and summarized as a "law." An irrational person's behavior follows no such pattern.

A third definition of rationality is the ability to understand, master, and coordinate all the variables necessary to create and manage an institution such as the economy, or society itself. This part of what Hayek called individualism false, the Rationalist tradition of Rene Descartes, the French Encyclopedists, and Rousseau. In *Discourse on Method*, Descartes expresses the belief that one person, if smart enough, can do the work of many. Hayek calls this "the fatal conceit" (Hayek 1988), the individual conceit which is fatal to everyone, perhaps everything other than the conceited. Hayek refers in particular to the institutions which form the backbone of Western society—the market, the family, the church. Hayek maintains, that these institutions are "the result of human action but not the result of human design" (Hayek 1948, p. 7), because no individual or group could have sufficient knowledge—sufficiently rational—to create them. In other words, rationality via individualism false is impossible, and socially undesirable because it leads to socialism and a loss of individual freedom. Human rationality is simply too limited relative to the complexity of the economy or society.

Individualism true, the "antirationalist" view, does not place great importance on the role of reason in human affairs. It views Man as irrational and not highly intelligent. Despite this humans can achieve a significant amount through freedom and the market process in which individuals know only about one or two things and which as a result they rely on many others, each of whom also knows only a relatively few things. He thus maintained that the most important aspect of knowledge is "unorganized," or tacit, what he called the knowledge of "particular circumstances of time and place." This knowledge is widely distributed among the population and can't be known or communicated by a central planning board. The tacit nature of knowledge is one example of Hayek's emphasis of the role of the unconscious in human life.

Hayek's philosophy of rationality is integrated with his writing on the human nervous system, expressed through his 1952 book *The Sensory Order*. Hayek's (1952) book, *The Sensory Order*, was an investigation into the relationship of the brain and memory, and the nature of the human mind. Hayek explained memory as the product of a pattern of brain neurons working together because of past experience. Hayek not only explained memory as a pattern of brain neurons, he then went on to make the analogy between these patterns, and the human mind, the market, and the limits to rationality. About 40 years after Hayek wrote *The Sensory Order* this became known as cortical memory networks. According to Joaquin Fuster, a neuroscientist at U.C.L.A.,

> The first proponent of cortical memory networks on a major scale was neither a neuro-scientist nor a computer scientist but, curiously, a Viennese economist: Frederick von Hayek... A man of exceptionally broad knowledge and profound insight into the operation of complex systems... Considering the neurobiological knowledge available even at the time of its publication, there is nothing dilettantish about that scholarly work... By postulating that all perception – and not just a part – is a product of memory, Hayek carries... one of the basic tenets of modern psychophysics. (Fuster 1999, p. 87)

Gerald Edelman, winner of the 1972 Nobel Prize in Physiology said of Hayk,

> Hayek made a quite fruitful suggestion, made contemporaneously by the psychologist Donald Hebb, that whatever kind of encounter the sensory system has with the world, a corresponding event between a particular cell in the brain and some other cell carrying the information from the outside word must result in reinforcement of the connection between those cells. These days, this is known as a Hebbian synapse, but von Hayek quite independently came upon the idea. I think the essence of his analysis still remains with us. (Edelman 1987, p. 25)

Just as Leibenstein did not believe that the determinant of the performance of a firm could be compared to an engineer reading a set of blueprints, Hayek did not believe that an individual "reads" the external environment the way an engineer reads a set of blueprints. Individuals do not perceive the "reality" of the world "out there." The mind interprets the incoming stimuli from the world. The sensory order, our experience of the external world, is our interpretation, of reality. The interpretation is due to the mind classifying all in-coming data from the external environment into an appropriate category based on the similarity of the in-coming data with all previously received data. Rizzello (1999) explains this process in his book *The Economics of the Mind*.

> Past *experience* plays a key role; it creates a system of links which records the situations when – in the history of the organism- the different groups of stimuli have worked together... Everything we perceive is immediately compared to other classes of recorded events. Every perception of a new stimulus or class of stimuli is influenced by the already existing classifications. A new phenomenon is always perceived in association with other events it shares something with. (Rizzello, p. 28)

The classification takes place in the unconscious mind which precedes sensation and perception. Therefore, the process of classification is neither conscious nor rational. Hayek says that "The process of experience thus does not begin with sensations or perceptions, but necessarily precedes them...experience is not a function of mind or consciousness, but that mind and consciousness are rather products of experience" (Hayek 1952, p. 166). Our current perception or experience of the world is, therefore, "path-dependent," based on past perceptions or experiences (Marciano 2009, p. 54).

Hayek spoke about the neural net, the human nervous system which includes the brain, and the spinal cord. The nervous system contains a very large number of neurons (nerve cells), estimated between 10 and 100 billion. Each neuron contains about 10,000 synapses. The number of connections among neurons is, therefore, between approximately 10^{15} to $5(10^{15})$—100 to 500 trillion connections. A rather long quote from *The Sensory Order* explains this part of Hayek's theory:

> ...the contention that the sensory (or mental) qualities are not in some manner originally attached to, or an original attribute of, the individual physiological impulses, but that the whole of these qualities is determined by the system of connexins by which the impulses can be transmitted from neuron to neuron; that it is thus the position of the individual impulse or group of impulses in the whole system of such connexins which gives it its distinctive quality; that this system of connexins is acquired in the course of the development of the species and the individual by a kind of 'experience' or 'learning'...we do not first have sensations which are then preserved by memory, but it is as a result of physiological memory that the physiological impulses are converted into sensations. The connexins between the physiological elements are thus the primary phenomenon which creates the mental phenomena. (Hayek 1952, p. 53)

New experiences activate different parts of the net. New pathways among neurons are created and these new pathways mean that we experience new knowledge. But the activation and the new pathways occur <u>before</u> the individual has a conscious experience. The unconscious process of classification precedes conscious experience. The human mind is the pattern of neuronal interactions, and when the pattern changes the mind changes. The mind is thus a *spontaneous order*.

Because the mental operations in the unconscious are so important in the process of conscious perception and thought Hayek prefers the name "super-conscious" to unconscious. The activity in the super-conscious is called abstractions while the activity in the conscious mind is called concrete. Hayek refers to the relationship between unconscious and conscious mental operations as "a casual one, that is, it refers to what, in the explanation of mental phenomena, must come first and can be used to explain the other... that these concrete particulars are the product of abstractions which the mind must possess in order that it should be able to experience particular sensations, perceptions, or images" (Hayek and Primacy pp. 36–37). Thus the term "the primacy of the abstract."

One implication is that "our actions must be conceived of as being guided by rules of which we are not conscious but which in their joint influence enable us to exercise extremely complicated skills without having an idea of the particular sequence of movements involved" (Primacy p, 38). There is thus a limit to how much the brain can explain. The complexity of the economy also led him to reject the idea of *homo economics* in favor of limited rationality and the use of customs and abstract, informal, unwritten or abstract rules whose origins no one knows but people follow.

The Persistence of the Complete Rationality Assumption

The First Generation questioned and presented arguments against the foundations of neoclassical theory. Research on the uses and effectiveness of heuristics (Gigerenzer 2016), and research on X-efficiency (Frantz 2016b) provided empirical support for the arguments. Historically the role of the First Generation will be seen as being the questioning and reconsidering the neoclassical model. The First Generation replaced the mechanical presentation of economics with a behavioral presentation.

Michael Jensen, well known for several seminal ideas, including the agency theory of the firm and the capital asset pricing model, argued in 2008 that perhaps 50% of our lives are ruled by something other than rational behavior. In "Non-Rational Behavior, Value Conflicts, Stakeholder Theory, and Firm Behavior," he says that

> human beings are not rational in something on the order of 50% of their lives… And by that I mean not only people out there in the world, but every single person, including me. The source of this nonrational behavior lies in the basic structure of the human brain. (Jensen 2008: 169)

In his 1987 article, "Rationality of the Self and Others in an Economic System," Kenneth Arrow argues against the monopoly of the rationality assumption in economics. He says that "Not only is it possible to devise complete models of the economy on hypotheses other than rationality, but in fact virtually every practical theory of macroeconomics is partly so based" (Arrow 1987: 202). Even more, he says that the rationality assumption is not essential to economics, and "the rationality hypothesis is by itself weak" (Arrow 1987: 206). It is most useful when markets are competitive, in equilibrium, and when they are "complete." Under other conditions "the very concept of rationality becomes threatened, because perception of others and, in particular, of their rationality becomes part of one's own rationality" (Arrow 1987: 203).

George Akerlof, 2001 winner of the Nobel Prize, and Janet Yellen, in their 1985 paper refer to this as "near" rationality. Near to rationality means that rationality is selective. In their paper they show that selective rationality has an effect on equilibrium solutions (Akerloff and Yellen 1985). Richard Thaler refers to selective rationality as "quasi" rationality meaning that an individual makes mistakes (Russell and Thaler 2001). They say that

> Since rationality is assumed, there is little in the literature to suggest what would happen if some agents were not rational. This is surprising in light of the accumulating evidence that supports Herbert Simon's view that man should be considered at most boundedly rational. (Russell and Thaler 2001: 1071)

Akerloff (1985), Arrow (1987), Thaler (2001), and Jensen (2008)—two Nobel Prize winners—add Herbert Simon (1957, 1976) and his theory of bounded rationality, and we have three Nobel Prize winners arguing against the notion of perfect rationality. Still, the assumption of complete rationality persists. One reason for the persistence is that economists do not have any theories about how we can predict behavior if people are not rational. Without rationality a person could do almost anything at any time and hence, for example, they may buy more at higher prices today but less tomorrow. They may borrow more money today at lower interest rates in the morning but less in the afternoon. The assumption of rationality means that human behavior conforms to a regular pattern, and hence serves as a "solid" basis for predictable human behavior. In other words, economists do economics and that ability depends on human behavior following a regular pattern.

There is I believe a second reason for the persistence of the rationality assumption. Near or quasi rationality is close to complete rationality without being completely rational. The welfare costs of near rationality are small so it should not be much of a concern. It can be discussed without causing too much worry that a concept so difficult to grasp or measure is being introduced into economic theory. The argument for quasi rationality is empirical support. But irrational meaning non-utility maximization behavior is difficult to identify, while at the same time it is "rarely fatal" (R&T, p. 1074). As with near rationality, discussing quasi rationality allows you to expand economic theory without too much concern that it will destroy one of the most useful concepts we possess. Akerloff and Thaler "dipped their toe" into the 'water' of reality without moving too far off 'neoclassical beach.' They were given credit for expanding economic theory without jeopardizing their reputations.

References

Akerloff, G., & Yellen, J. (1985). A near-rational model of the business cycle, with wage and price inertia. *Quarterly Journal of Economics, 100*(Supplement), 823–838.

Arrow, K. (1987). Rationality of self and others in an economic system. In R. Hogarth & M. Reder (Eds.), *Rational choice: The contrast between economics and psychology* (pp. 201–216). Chicago: University of Chicago Press.

Edelman, G. (1987). *Neural Darwinism*. New York: Basic books.

Frantz, R. (Ed.). (2007). *Renaissance in behavioral economics: Essays in honour of Harvey Leibenstein*. London: Routledge.

Frantz, R. (2013). Hayek and behavioral economics. In R. Frantz, & R. Leason (Eds.), *Houndmill*. U.K.: Palgrave Macmillan.

Frantz, R. (2016a). *Minds, Markets, and Mileaux. Commemorating the Centennial of the Birth of Herbert Simon*. In R. Frantz & L. M. Houndmill (Eds.),. U.K.: Palgrave Macmillan.

Frantz, R. (2016b). 50 Years of X-efficiency research. *Working Paper*.

Frantz, R. (2017). *Handbook of behavioral economics*. In R. Frantz, S.-H. Chen, K. Dopfer, F. Heukelom & S. Mousavi, (Eds.),. New York: Routledge.

Fuster, J. (1999). *Memory in the Cerebral Cortex*. Cambrisge: MIT Press.

Gigerenzer, G. (2016). Towards a Rational Theory of Heuristics. In R. Frantz & L. Marsh (Eds.), *Minds, Markets, and Milieux. Commemorating the Centennial of the Birth of Herbert Simon*, pp. 34–59. London: Palgrave Macmillan.

Hayek, F. (1948). "Individualism true and false." In F. Hayek (Ed.), *Individualism and economic order*, pp. 1–32. Chicago: University Chicago Press.

Hayek, F. (1952). *The Sensory Order. Abingdon on Thames*. UK: Routledge, Kegan Paul PLC.

Hayek, F. (1967). The Theory of Complex Phenomena. In F. A. Hayek (Ed.), *Studies in Philosophy, Politics and Economics* (pp. 22–42). Chicago: University of Chicago Press.

Hayek, F. (1988). *The Fatal Conceit. The Errors of Socialism*. Chicago: University of Chicago Press.

Jensen, M. (2008). Non-rational behavior, value conflicts, stakeholder theory, and firm behavior. *Business Ethics Quarterly, 18*(2), 167–171.

Katona, G. (1940). *Organizing and Memorizing*. New York: Columbia University Press.

Katona, G. (1960). *The Powerful Consumer*. New York: McGraw-Hill.

Katona, G. (1963). *Psychological Analysis of Economic Behavior*. New York: McGraw Hill.

Katona, G. (1975a). *Psychological Economics*. New York: Elsevier.

Katona, G. (1975b). *Psychological Economics*. New York: Elsevier.

Katona, G. (1980). *Essays in Behavioral Economics*. Ann Arbor: Survey Research Center. Institute for Social Research, Univ Michigan.

Leibenstein, H. (1966). Allocative Efficiency and X-Efficiency. *American Economic Review, 56* (June), 392–415.

Leibenstein, H. (1976). *Beyond Economic Man*. Cambridge, MA.: Harvard University Press.

Leibenstein, H. (1987). *Inside the Firm. The Inefficiencies of Hierarchy*. Cambridge, MA.: Harvard University Press.

Marciano, A. (2009). "Why Hayek is an Darwinian (After All)? Hayek and Darwin on Social Evolution." *Journal of Economic Behavior and Organization, 71*, July: 52–61.

Mundell, R. (1962). "Review of L. H. Janssen, Free Trade, Protection and Customs Union." *American Economic Review, 52*, June: 622.

Pareto, V. (1971). *Manual of Political Economy*. New York: Augustus M. Kelley.

Rizello, S. (1999). *The Economics of the Mind*. Northampton: E.E. Elgar.

Robbins, L. (1932). *An Essay on the Nature and Significance of Economic Science*. New York: Macmillan.

Russell, T., & Thaler, Richard. (2001). The Relevance of Quasi Rationality in Competitive Markets. *American Economic Review, 75*(5), 1071–1082.

Samuelson, P. (1970). "Maximum Principles in Analytical Economics." The Nobel Foundation.

Simon, H. (1955). "A behavioral model of rational choice." *Quarterly Journal of Economics, 69* (1), 99–118.

Simon, H. (1957). *Models of Man*. NY: John Wiley.

Simon, H. (1976). From Substantive to Procedural Rationality. In S. J. Latsis (ed.), *Method and Appraisal in Economics*, pp. 129–148. Cambridge: Cambridge Univ Press.

Simon, H. (1959). Theories of decision-making in economics and behavioral science. *American Economic Review, 49*(3), 253–83.

Simon, H. (1982a). *Models of bounded rationality*. In *Empirically grounded economic reasoning*, Vol. 3. Cambridge: MIT Press.

Simon, H. (1982b). "A behavioral model of rational choice." In H. Simon (Ed.), *Models of bounded rationality. Behavioral economics and business organiza tion* (Vol. 2, pp. 239–258). Cambridge, MA.

Simon, H. (1982c). "From substantive to procedural rationality." In H. Simon (Ed.), *Models of bounded rationality. Behavioral economics and business organization,* (vol. 2, pp. 424–443). Cambridge, MA: MIT Press.

Simon, H. (1983). *Reason in Human Affairs*. Palo Alto: Stanford Univ Press.

Simon, H. (1996). *The Sciences of the Artificial* (2nd ed.). Cambridge, MA: MIT Press.

Simon, H. (1997). *Administrative Behavior* (4th ed.). New York: Free Press.

Chapter 7
Embodied Rationality

Shaun Gallagher

Abstract Recent developments in embodied cognition in the field of cognitive science support an expanded notion of rationality. I attempt to explicate this expanded notion by introducing the concepts of embodied rationality and enactive hermeneutics. I argue that bodily performance is rational and that there is continuity between the rational movements of the body and reflective thinking understood as a skill.

Introduction

The well-known existential philosopher, Gabriel Marcel, proposed a distinction between the concepts of mystery and problem. Marcel defined a problem as something that could be identified as objectively distinct from the subject or agent—something that one could get a perspective on from a distance. One may be able to solve problems using science or instrumental rationality. In contrast, Marcel defined a mystery as something that so involves the subject or agent that gaining an objective perspective on it is not possible. It is not something that can be solved by scientific or instrumental rationality, not because it's a difficult problem, but because it is something different from a problem. The defining feature of a mystery is the inextricable involvement of the existential subject.

> A problem is something which I meet, which I find completely before me, but which I can therefore lay siege to and reduce. But a mystery is something in which I am myself involved, and it can therefore only be thought of as a sphere where the distinction between what is in me and what is before me loses its meaning and initial validity. (Marcel 1949, 117)

S. Gallagher (✉)
University of Memphis, Memphis, USA
e-mail: s.gallagher@memphis.edu

S. Gallagher
University of Wollongong, Wollongong, Australia

© Springer International Publishing AG, part of Springer Nature 2018
G. Bronner and F. Di Iorio (eds.), *The Mystery of Rationality*,
https://doi.org/10.1007/978-3-319-94028-1_7

Marcel suggested that being in love is a good example of a mystery. But also, one's own existence, and one's own body (not the body as a biological object, but the body as it is lived or experienced by the embodied subject) are mysteries in this sense.

Marcel is not the only philosopher to suggest such a distinction. Noam Chomsky made a very similar one with respect to the study of language and the mind, indeed, using the same terminology (1976, 281).

> I would like to distinguish roughly between two kinds of issues that arise in the study of language and mind: those that appear to be within the reach of approaches and concepts that are moderately well understood — what I will call "problems"; and others that remain as obscure to us today as when they were originally formulated — what I will call "mysteries."

Although Gadamer (2004, 301) did not use the term 'mystery', his characterization of the 'hermeneutical situation' involved a similar structure to what Marcel called a mystery—the inextricable involvement of the interpreter. Likewise, Dewey (1938) defined the concept of 'situation', not as equivalent to the external environment, but as inclusive of the agent (or the organism) as it is coupled to the environment. The distinction between the 'geographical environment' and the 'behavioral environment' (Koffka 2013, 27–51) is similar in so far as the latter concept includes the comporting agent. Along these lines, and I think in the same spirit, Aristotle (350 BCEa, I, 3) famously noted that although in many subject areas problems could be addressed by theoretical or scientific knowledge, in the area of ethics, where human action is at stake, issues are not as well ordered, and one would have to be satisfied with a precision 'just so far as the nature of the subject admits'.

Faced with these different suggestions and distinctions, there are at least two possible stances. Either one takes the 'mystery' (to stay with Marcel's term) to be irrational; or one widens the definition of rationality to include mystery. Aristotle's notion of *phronesis* (practical wisdom), Dewey's pragmatism, Gadamer's hermeneutics, and Marcel's existentialism are attempts to expand the notion of rationality beyond the instrumental or strict natural scientific conceptions. I'll argue in this paper that recent developments in embodied cognition in the field of cognitive science also support this expanded notion of rationality. I'll attempt to explicate this expanded notion by introducing the concepts of embodied rationality and enactive hermeneutics.

Enactivist Approaches to Embodied Cognition

There are a number of different approaches to embodied (or situated) cognition (EC). Some approaches attempt to remain close to the standard cognitivist conception of the mind based on computational models and mental representations. One such approach, so-called 'weak' EC (Alsmith and Vignemont 2012), is worked out in terms of internal body (B-) formatted representations. Goldman (2014), for example, takes the role of the body to be fully accounted for in terms of neural processes in which sensory-motor representations get reiterated for purposes other than sensory-motor function. For example, B-formatted simulations may be important for language and concept processing (e.g., Glenberg 2010; Meteyard et al. 2012; Pezzulo et al. 2011; Pulvermüller 2005). Extended functionalist approaches, like the extended mind hypothesis (Clark and Chalmers 1998; Menary 2010), also remain tied to computational rationality even as they minimize the role of mental representations. On this model, some of the work of the mind is taken up through an embodied engagement with tools, instruments and artifacts in the environment.

A more radical break with computational models is made by enactivist approaches, which call for a radical change in the way we think about the mind and brain. These approaches emphasize dynamical, non-linear self-organizing systems that constitute cognition from the bottom up, where cognition is defined in terms of responses made by autopoietic biological organisms to environmental conditions (Thompson 2007; Varela et al. 1991). Human cognition obviously involves the brain, but the brain is considered part of the larger system of brain-body-environment. In this dynamical intertwinement of brain and body with the environment, the brain is not computing across representations. It's not representing or creating an internal model of the world. It's not taking the world as a problem to be solved.[1] It's rather contributing to the ongoing responses made by the embodied agent to physical, social and cultural environments, in ways that allow for the enactment of meaning.

The enactivist approach can be summarized by the following background assumptions.[2]

(1) Cognition is not simply a brain event. It emerges from processes distributed across brain-body-environment;
(2) The world (meaning, intentionality) is not pre-given or predefined, but is structured by cognition and action;
(3) Cognitive processes acquire meaning in part by their role in the context of action, rather than through a representational mapping or replicated internal model of the world;
(4) The enactivist approach has strong links to dynamical systems theory, emphasizing the relevance of dynamical coupling and coordination across brain-body-environment;
(5) In contrast to classic cognitive science, which is often characterized by methodological individualism with a focus on internal mechanisms, the

---

[1]Enactivism, however, may be viewed as consistent with predictive models of brain function as long as the relation between brain, body and environment is properly conceived (see Gallagher and Allen 2016).

[2]See Gallagher (2017). These assumptions are drawn from the following sources: Clark (1999), Di Paolo et al. (2010), Dominey et al. (2016), Engel (2010), Engel et al. (2013), Thompson and Varela (2001), Varela et al. (1991).

enactivist approach emphasizes the extended, intersubjective and socially situated nature of cognitive systems;

(6) Enactivism aims to ground higher and more complex cognitive functions not only in sensorimotor coordination, but also in affective and autonomic aspects of the full body;

(7) Higher-order cognitive functions, such as reflective thinking or deliberation are exercises of skillful know-how and are usually coupled with situated and embodied actions.

Enactivist versions of EC emphasize the idea that perception is *for action*, and that action-orientation shapes most cognitive processes. Actions and bodily responses are not mindless, however. Some claims by Dreyfus (2007), a proponent of a version of enactivist EC, might be taken as suggesting that the organism's intelligent coping with its environment is in some way mindless. In a recent debate with John McDowell, for example, he rejects what he calls the 'myth of the mental', and holds that perception and action most often occur without mental intervention. McDowell (2007), in response, argues that perception (and agency) and embodied coping are conceptual and rational, and therefore not as 'mindless' as Dreyfus contends. McDowell explains, however, that rationality does not have to be situation independent, and this can be seen in the Aristotelian notion of *phronesis* as a model for situated rationality—one that Dreyfus himself takes as a model for embodied coping. For McDowell, *phronesis* involves an initiation into conceptual capacities. Dreyfus (2005, 51), however, takes *phronesis* to be 'a kind of understanding that makes possible an immediate response to the full concrete situation'. McDowell, influenced by Gadamer on this point, contends that 'the practical rationality of the *phronimos* [the person with practical reason] is displayed in what he does even if he does not decide to do that as a result of [explicit, deliberative] reasoning' (2007, 341). Rationality is built into action insofar as we can think of reasoning as the ability to differentiate which affordances to respond to, and how to go about responding to them. McDowell argues that the fact that we are able to give reasons for our action, even if we did not form deliberative reasons prior to the action, suggests that our actions and embodied copings have an implicit structure that is rational (or proto-rational) and amenable to conceptuality.

This question about the nature of rationality becomes central to the debate between Dreyfus and McDowell. On the one hand, for Dreyfus, the concept of rationality is not something inherent in life or action. In this regard, he thinks of rationality in terms of giving reasons for our actions, which involves detached, reflective thoughtful processes associated with language—propositional discourse, the space of reasons, or conceptual articulation. On the other hand, for McDowell, rational language use is closely tied to the situation in which it occurs. Our openness to the world involves a situated categorial aspect, which allows us to register it linguistically (even if we don't always do so). Even if we are not 'ready in advance' to put a word to every aspect of experience, we have an anticipatory understanding that conceptually shapes our experience, and allows us to "step back" to reflectively identify the rationality of our actions. To this Dreyfus responds:

I agree with McDowell that we have a freedom to step back and reflect that nonhuman animals lack, but I don't think this is our most pervasive and important kind of freedom. Such stepping back is intermittent in our lives and, in so far as we take up such a 'free, distanced orientation', we are no longer able to act in the world. I grant that, when we are absorbed in everyday skillful coping, we have the capacity to step back and reflect but I think it should be obvious that we cannot exercise that capacity without disrupting our coping. (Dreyfus 2007, 354).

Here the notion of affordance (Gibson 1977) offers some explanatory help. The concept of affordance is defined as having the same relational structure as Dewey's notion of 'situation'. That is, it does not signify something about an objective structure in the environment; it's defined in terms of the *relation* between a certain type of organism or agent, which has a certain skill level, and the particularities of the environment. Dreyfus distinguishes affordances as facts (or as Rietveld and Kiverstein 2014 put it, affordances relative to a way of life) from affordances as specific solicitations, here and now. 'Although when we step back and contemplate them affordances can be experienced as features of the world, when we respond to their solicitations they aren't figuring for a subject as features of the world [in McDowell's sense]' (2007, 358). McDowell, according to Dreyfus, assumes that the world is already a set of facts that are determinate and that can then be named, and thought, and fit into rational concepts. In contrast, for Dreyfus, the world is indeterminate, 'not implicitly conceptual and simply waiting to be named. Our relation to the world is more basic than our mind's being open to apperceiving categorially unified facts' (2007, 359). This fits well with the enactivist idea that, rather than finding meaning pre-formed, we enact meaning in our ongoing actions.

Embodied Rationality

The notion of embodied rationality splits the difference between Dreyfus and McDowell. On the enactivist view, the world is laid out in perception, not in terms of a conceptual, or proto-conceptual meaning, but first of all, in terms of differentiations that concern my action possibilities or affordances—the object is something I can reach, or not; something I can lift, or not; something I can move or not. One's ability to make sense out of the world comes, *in part,* from an active and pragmatic engagement with the world. If we can then turn around or step back to discover that our world or our experience has an inherent rational or proto-conceptual structure, that's because that structure has already been put there by our pre-predicative embodied engagements. Dreyfus may reject this as a form of rationality because he is thinking of rationality in the standard way. As Zahavi (2013) points out, both Dreyfus and McDowell continue to retain and share an overly intellectualized (conceptualized, "languaged") conception of the mind. In contrast to this traditional conception (which is the concept of mind that Dreyfus rejects and McDowell accepts), the alternative is to think of mental skills such as reflection, problem solving, decision making, and so on, as enactive, non-representational forms of embodied coping that emerge from a pre-predicative perceptual ordering of differentiations and similarities (Gallagher 2017).

Consider, that there is a rationality that is implicit in the hand. This kind of hand-related rationality has been recognized in a long tradition of philosophy that goes back to Anaxagoras' observation (located on the Dreyfus side of the debate) that we humans are the wisest of all beings because we have hands—human rationality derives from human practices. Aristotle turns that claim around to make it less enactivist and more consistent with McDowell's view: "Man has hands because he is the wisest of all beings"—human practices derive from human rationality.

> Now it is the opinion of Anaxagoras that the possession of these hands is the cause of man being of all animals the most intelligent. But it is more rational to suppose that his endowment with hands is the consequence rather than the cause of his superior intelligence. For the hands are instruments or organs, and the invariable plan of nature in distributing the organs is to give each to such animal as can make use of it.... We must conclude that man does not owe his superior intelligence to his hands, but his hands to his superior intelligence. (Aristotle 350b)

Still, in the Aristotelian tradition the hand is raised to the level of the rational by considering it the *organum organorum*.[3] For the enactivist, hands (as part of a complete system that includes the brain and the rest of the body) are action oriented. As an agent reaches to grasp something, the hand automatically (and without the agent's conscious awareness) shapes itself into just the right posture to form the most appropriate grip for that object and for the agent's purpose. If I reach to grab a banana in order to take a bite, the shape of my grasp is different from when I reach to grab a banana in order to pretend it's a phone. Differences in my grasp reflect my intention so that if I grasp the fruit to eat it, the kinematics of my movement are different from when I grasp it to offer it to you, and different again from when I grasp it to throw at you (Ansuini et al. 2006, 2008; Jeannerod 1997; Marteniuk et al. 1987; Sartori et al. 2011). Hands are integrated with visual perception (via the dorsal visual pathway), so that I see the fruit as graspable for specific purposes. The brain evolved to do what it does in this regard only because it had hands to work with—hands that evolved with the brain in a holistic relation with other bodily aspects.

It is sometimes the case that very smart hand-brain dynamics take the lead over a more conceptual, ideational intelligence. For example, a patient with visual agnosia who is unable to recognize objects, when shown a picture of a clarinet, calls it a 'pencil'. At the same time, however, his fingers began to play an imaginary clarinet (Robertson and Treisman 2010, 308). The hands, and more generally, the body and its movement in this regard are rational and perform a kind of 'manual thinking' (Bredekamp 2007) that integrates its action across all perceptual modalities. Thus

---

[3]Beyond the Aristotelian view, Newton suggested that the thumb is good evidence of God's existence (see Dickens 1908, 346), and Kant (1992) used his hands (the fact that hands are incongruent counterparts, e.g. a left hand doesn't fit properly into a right-hand glove) to prove that Newton was right about space being absolute.

we have a rationality involved in touch and haptic exploration, hand-mouth coordination, hand-eye coordination, shading the eyes; cupping one's ears, or holding one's nose.

The hand facilitates perception and action; it also transforms its movements into language (via gesture) and into thinking. Empirical studies suggest that there are close relationships between gesture, speech, and thinking—they are part of the same system, which David McNeill calls the hand-language-thought system (Cole et al. 2002; McNeill et al. 2008; Quaeghebeur et al. 2014). Gesture adds to cognitive ability, and does so without requiring distance, discontinuity, or a 'stepping back' that comes between this kind of movement and spoken language—nor between manual thinking and thinking proper—they are part of the same performance system.

In this precise sense, bodily performance is rational and demonstrates a continuity between the rational movements of the body and reflective thinking. Reflective thinking is a *skill* as much as physical coping is. Even if it is a step back from action, it is itself a form of action or performance. Reflection need not be disconnected from an expert performance, but can be integrated as part of that performance—a dimension of the flow rather than something different from it. It could be the type of reflection that can occur during musical performance even as the musician stays in the flow (Høffding 2015; Salice, Høffding & Gallagher 2017). In some circumstances the expertise of the teacher or the musician is just this ability to do both at once, and it would be odd to claim that this kind of performance is not expert performance because one is able to reflect as one is engaged in the performance. Such reflections are likely different in each case, but nonetheless nuanced and integrated with physical actions.[4]

Things in the environment that count as, and that we perceive as, salient or significant affordances are laid out along affective, hedonic lines that are tied to other agents and their actions. Our perception of objects is shaped not simply by bodily pragmatic or enactive possibilities, but also by a certain intersubjective saliency that derives from the behavior and emotional attitude of others. Both Dreyfus and McDowell make use of the concept of *phronesis* as a practiced excellence in knowing what to do. *Phronesis* is closely tied to the particularities of each situation. It can be intuitive/automatic (Dreyfus) *and/or* it can involve reflective/deliberative skills (McDowell). Most importantly, however, *phronesis* is

---

[4]An anonymous reviewer suggested that "we have still to give a place in this very framework to abstraction, generalization, theory and formalization as useful tools—for planning coordinated actions, for extending previous methods to new situations and new domains, for making revisions in order to restore coherence in a belief base, etc." I agree. We can argue for a kind of continuity between embodied practices and reflective practices even if the former are considered more concrete than the latter. We do not have to conceive of the continuity as hierarchically organized. Consider the suggestion made by Goldstein and Scheerer: "Although the normal person's behaviour is prevailingly concrete, this concreteness can be considered normal only as long as it is embedded in and codetermined by the abstract attitude. For instance, in the normal person both attitudes are always present in a definite figure-ground relation" (Goldstein and Scheerer 1964, 8; see Gallagher (2017), for more on this gestalt view of the relations between embodied and reflective practices.

intersubjective. As Aristotle tells us, *phronesis* is something we learn through being with others. In developmental terms, we learn it in very basic, intersubjective interactions—seeing things as others see them, imitating, doing what others do, valuing what others value—in processes that involve embodied rationality, social norms, situated reflection, etc. *Phronesis*, even if it sometimes involves situated reflective thinking, is continuous with and cut from the same fabric as embodied coping, which involves interaction as much as action.

Enactivist Hermeneutics

On the enactivist EC view, the world (meaning, intentionality) is not pre-given or predefined, but is structured within the dynamical processes of situated action, and shows up as an affordance space (Brincker 2014).[5] An affordance space consists of the range of possibilities provided by the dynamical relations between body and environment. A change in affordance space is generated by any change in this relation. 'An individual's occurrent affordance space is defined by evolution ([e.g.,] the fact that she has hands), development (her life-stage—infant, adult, aged), and by social and cultural practices (and their normative constraints)—all of which enable and constrain the individual's action possibilities' (Gallagher 2015, 342). In evolutionary terms, the human affordance space differs from a non-human animal's, defined, for example by the fact that humans have hands and capacities for specific kinds of movement. A child's affordance space differs from an adult's due to differences in development. Humans learn to move or think in specific ways across ontogenetic parameters. Likewise, one's individual affordance space differs from another's due to differences in prior experience, skill level, education and cultural practices, etc. Across these parameters humans are enabled or constrained to move and to think in particular ways due to their prior experiences, and due to plastic and 'metaplastic' (Malafouris 2013) changes in both brain and body, but also within the constraints of their environment.

If, as Merleau-Ponty suggests, traditional conceptions of rationality presuppose "a universe perfectly explicit in itself" (2012, 44), embodied rationalism, in contrast, lives with ambiguity—the kind of ambiguity that comes from the fact that we cannot extricate ourselves from the situation, but rather, as embodied, we are necessarily involved in an intentional arc that cuts across brain, body, and environment. Ambiguity here means an incompleteness or lack of coincidence that describes the situated first-person perspective. Again, similar to Marcel's conception of mystery, Dewey's notion of situation captures the point. 'In actual experience, there is never any such isolated singular object or event; an object or event is always a special part, phase, or aspect, of an environing experienced world—a

---

[5]Rietveld and Kiverstein (2014) use the term 'landscape of affordances' and define it relative to a form of life.

situation' (Dewey 1938, 67). For Dewey, it is not the organism that is placed in a situation. Rather the situation is constituted by organism-environment, which means that the situation already includes the agent or experiencing subject. In this regard, for example, I cannot strictly point to the situation because my pointing is part of the situation. I cannot speak of it as some kind of distanced objective set of facts because my speaking is part of it. My movement is a movement of the situation. My reflection, and even my stepping back, is likewise part of the situation. Situated coping does not mean simply to rearrange objects in the environment, but to rearrange oneself as well. Indeed, any adjustment one makes to objects, artifacts, tools, practices, social relations or institutions, is equally an adjustment one makes to oneself.

An enactivist hermeneutics is, accordingly, a hermeneutics of affordances and ambiguities. An agent interprets the world in terms of what Husserl (1989) called the 'I can'. The 'I can' expresses the idea that agents perceive their world in pragmatic terms of what they can do with things, or how they can interact with others. Another way to say this is that an enactivist hermeneutics involves the acknowledgement that interpretation is always *in* and *of* a hermeneutical *situation* where 'situation' (as both Dewey and Gadamer suggested) includes not only the historical and horizonal limits of interpretation, but the interpreter him- or herself, understood as an agent, and the affordances (physical, pragmatic, social and cultural) that are relative to that agent.

> The very idea of a situation means that we are not standing outside of it; hence we are unable to have any [purely] objective knowledge concerning it. We always find ourselves within a situation, and throwing light on it is a task that is never entirely finished. This is also true of the hermeneutic situation (Gadamer 2004, 301)

What then is rationality for an enactivist hermeneutics in which body and environment are co-relational; and in which this co-relationality shapes our interpretations of what things mean and what value they may have? It's not an observational or spectatorial stepping back that detaches from the situation to frame the world in abstract concepts. This latter form of rationality, traditionally characterized as the primary or highest form of knowledge—theoretical knowledge (as in Aristotle), or scientific knowledge (as in the positivists)—may in fact be derived or secondary, and limited. Limited insofar as, even if it does aspire to encompass, epistemologically or formally, everything there is (as in Leibniz's conception of divine rationality[6]), and to do so from an objective (view-from-nowhere) perspective, it fails to capture the

---

[6]'God is that which perceives perfectly whatever can be perceived' (Leibniz, cited in Nachtomy 2008, 74). Nachtomy (2008, 80) summarizes Leibniz's 'hard' conception of rationality as a 'divine combinatorics': 'The combinatorial nature of concepts serves as the formal and universal structure of all concepts by stipulating a calculus of all the consistent combinations among all simple forms in God's mind. The combinatorial nature of concepts applies to human thought as well. However, humans must substitute the variables – and the simple elements and the combinatorial rules – with notations, including the "alphabet" and the syntax of actual sciences, practices, and applications, such as written languages, geometry, music, chemistry....' See the related notion of a *mathesis universalis*.

ambiguity intrinsic to the relational dynamics of the agentive situation; and secondary because it may be the product or outcome of abstracting from and internalizing the more basic (pragmatic, and in Marcel's sense, 'mysterious') rationality that characterizes the worldly engagements of participants in embodied situations, and operates as its shaky and imperfect ground.

Acknowledgements Research on this paper was supported by the Australian Research Council Discovery Project "Minds in Skilled Performance" (DP170102987), and by the Humboldt Foundation's Anneliese Maier Research Award.

References

Alsmith, A. J. T., & de Vignemont, F. (2012). Embodying the mind and representing the body. *Review of Philosophy and Psychology, 3*(1), 1–13.

Ansuini, C., Giosa, L., Turella, L., Altoè, G. M., & Castiello, U. (2008). An object for an action, the same object for other actions: Effects on hand shaping. *Experimental Brain Research, 185,* 111–119.

Ansuini, C., Santello, M., Massaccesi, S., & Castiello, U. (2006). Effects of end-goal on hand shaping. *Journal of Neurophysiology, 95,* 2456–2465.

Aristotle. (350 BCEa). *Nicomachean ethics.* Trans. W. D. Ross. http://classics.mit.edu/Aristotle/nicomachaen.html.

Aristotle. (350 BCEb). *On the parts of animals.* Trans. W. Ogle. http://classics.mit.edu/Aristotle/parts_animals.html.

Bredekamp, H. (2007). *Galilei der Künstler. Die Zeichnung, der Mond, die Sonne.* Akademie-Verlag.

Brincker, M. (2014). Navigating beyond 'here & now' affordances—On sensorimotor maturation and 'false belief' performance. *Frontiers in Psychology, 5,* 1433. https://doi.org/10.3389/fpsyg.2014.01433.

Chomsky, N. (1976). Problems and mysteries in the study of human language. In *Language in focus: Foundations, methods and systems* (pp. 281–357). Netherlands: Springer.

Clark, A. (1999). An embodied cognitive science? *Trends in Cognitive Sciences, 3*(9), 345–351.

Clark, A., & Chalmers, D. (1998). The extended mind. *Analysis, 58*(1), 7–19.

Cole, J., Gallagher, S., & McNeill, D. (2002). Gesture following deafferentation: A phenomenologically informed experimental study. *Phenomenology and the Cognitive Sciences, 1*(1), 49–67.

Dewey, J. (1938). *Logic: The theory of inquiry.* New York: Holt, Rinehart & Winston.

Dickens, C. (1908). *Miscellaneous Papers: From the Morning Chronicle, the Daily News, the Examiner, All the Year Round, and Other Sources.* London: Chapman & Hall.

Di Paolo, E. A., Rohde, M., & De Jaegher, H. (2010). Horizons for the enactive mind: Values, social interaction, and play. In J. Stewart, O. Gapenne, & E. A. Di Paolo (Eds.), *Enaction: Toward a new paradigm for cognitive science* (pp. 33–87). Cambridge, MA: MIT Press.

Dominey, P. F., Prescott, T., Bohg, J., Engel, A.K., Gallagher, S., & Heed, T., et al. (2016). Implications of action-oriented paradigm shifts in cognitive science. In *Where's the action? The pragmatic turn in cognitive science.* Cambridge, MA: MIT Press.

Dreyfus, H. L. (2005). Overcoming the myth of the mental: How philosophers can profit from the phenomenology of everyday expertise. In *Proceedings and Addresses of the American Philosophical Association* (pp. 47–65). American Philosophical Association.

Dreyfus, H. L. (2007). Why Heideggerian AI failed and how fixing it would require making it more Heideggerian. *Philosophical Psychology, 20*(2), 247–268.

Engel, A. K. (2010). Directive minds: How dynamics shapes cognition. In J. Stewart, O. Gapenne, & E. Di Paolo (Eds.), *Enaction: Towards a new paradigm for cognitive science* (pp. 219–243). Cambridge: MIT Press.

Engel, A. K., Maye, A., Kurthen, M., & König, P. (2013). Where's the action? The pragmatic turn in cognitive science. *Trends in Cognitive Sciences, 17*(5), 202–209.

Gadamer, H. G. (2004). *Truth and method.* Trans. J. Weinsheimer & D. G. Marshall. New York: Bloomsbury Publishing.

Gallagher, S. (2017). *Enactivist interventions: Rethinking the mind.* Oxford: Oxford University Press.

Gallagher, S. (2015). Doing the math: Calculating the role of evolution and enculturation in the origins of mathematical reasoning. *Progress in Biophysics and Molecular Biology, 119,* 341–346.

Gallagher, S., & Allen, M. (2016). Active inference, enactivism and the hermeneutics of social cognition. *Synthese.* https://doi.org/10.1007/s11229-016-1269-8.

Gibson, J. J. (1977). The theory of affordances. In R. Shaw & J. Bransford (Eds.), *Perceiving, acting, and knowing* (pp. 67–82). Hillsdale, NJ: Lawrence Erlbaum.

Glenberg, A. M. (2010). Embodiment as a unifying perspective for psychology. *Wiley Interdisciplinary Reviews: Cognitive Science, 1*(4), 586–596.

Goldman, A. I. (2014). The bodily formats approach to embodied cognition. In U. Kriegel (Ed.), *Current controversies in philosophy of mind* (pp. 91–108). New York and London: Routledge.

Goldstein, K., & Scheerer, M. (1964). *Abstract and concrete behavior. An experimental study with special tests.* Evanston, IL: Northwestern University. Reprint of *Psychological Monographs* 53 (2), 1941.

Høffding, S. (2015). A phenomenology of expert musicianship. PhD Dissertation, Department of Philosophy, University of Copenhagen.

Husserl, E. (1989). *Ideas pertaining to a pure phenomenology and to a phenomenological philosophy, second book.* Trans. R. Rojcewicz and A. Schuwer. Dordrecht: Kluwer Academic Publishers.

Jeannerod, M. (1997). *The cognitive neuroscience of action.* Oxford: Blackwell Publishers.

Kant, I. (1992). Concerning the ultimate ground of the differentiation of directions in space. In D. Walford & R. Meerbote (Eds.), *The Cambridge Edition of the Works of Immanuel Kant. Theoretical Philosophy, 1755–1770* (pp. 365–372). Cambridge: Cambridge University Press.

Koffka, K. (2013). *Principles of Gestalt psychology.* London: Routledge.

Malafouris, L. (2013). *How things shape the mind.* Cambridge, MA: MIT Press.

Marcel, G. (1949). *Being and having.* Trans. K. Farrer. Westminster, UK: Dacre Press.

Marteniuk, R. G., MacKenzie, C. L., Jeannerod, M., Athenes, S., & Dugas, C. (1987). Constraints on human arm movement trajectories. *Canadian Journal of Psychology, 41,* 365–378.

McDowell, J. (2007). What myth? *Inquiry, 50*(4), 338–351.

McNeill, D., Duncan, S., Cole, J., Gallagher, S., & Bertenthal, B. (2008). Neither or both: Growth points from the very beginning. *Interaction Studies, 9*(1), 117–132.

Menary, R. (Ed.). (2010). *The extended mind.* Cambridge, MA: MIT Press.

Merleau-Ponty, M. (2012). *Phenomenology of perception.* Trans. D. A. Landes. London: Routledge.

Meteyard, L., Cuadrado, S. R., Bahrami, B., & Vigliocco, G. (2012). Coming of age: A review of embodiment and the neuroscience of semantics. *Cortex, 48*(7), 788–804.

Nachtomy, O. (2008). Leibniz's rationality: Divine intelligibility and human intelligibility. In M. Dascal (Ed.), *Leibniz: What kind of rationalist?* (pp. 73–82). Dordrecht: Springer.

Pezzulo, G., Barsalou, L. W., Cangelosi, A., Fischer, M. H., McRae, K., & Spivey, M. J. (2011). The mechanics of embodiment: a dialog on embodiment and computational modeling. In A. Borghi & D. Pecher (Eds.), *Embodied and grounded cognition* (p. 196). Frontiers E-books.

Pulvermüller, F. (2005). Brain mechanisms linking language and action. *Nature Reviews Neuroscience, 6,* 576–582.

Quaeghebeur, L., Duncan, S., Gallagher, S., Cole, J., & McNeill, D. (2014). Aproprioception and gesture. In C. Müller, E. Fricke, A. Cienki, S. H. Ladewig & D. McNeill (Eds.), *Handbook on body—language—communication* (2048–2061). De Gruyter-Mouton Publisher.

Rietveld, E., & Kiverstein, J. (2014). A rich landscape of affordances. *Ecological Psychology, 26* (4), 325–352.

Robertson, L. C., & Treisman, A. (2010). Consciousness: Disorders. In E. B. Goldstein (Ed.), *Encyclopedia of perception*. New York: Sage.

Salice, A., Høffding, S., & Gallagher, S. (2017). Putting plural self-awareness into practice. *Topoi*. https://doi.org/10.1007/s11245-017-9451-2.

Sartori, L., Becchio, C., & Castiello, U. (2011). Cues to intention: The role of movement information. *Cognition, 119,* 242–252.

Thompson, E. (2007). *Mind in life: Biology, phenomenology and the sciences of mind*. Cambridge, MA: Harvard University Press.

Thompson, E., & Varela, F. (2001). Radical embodiment: Neural dynamics and consciousness. *Trends in Cognitive Sciences, 5*(10), 418–425.

Varela, F. J., Thompson, E., & Rosch, E. (1991). *The embodied mind: Cognitive science and human experience*. Cambridge: MIT Press.

Zahavi, D. (2013). Mindedness, mindlessness and first-person authority. In J. K. Schear (Ed.), *Mind, reason, and being-in-the-world: The McDowell-Dreyfus debate* (pp. 320–340). London: Routledge.

Chapter 8
Rational Choice Explained and Defended

Herbert Gintis

Abstract Choice behavior can generally be best modeled using the rational actor model, according to which individuals have a time-, state-, and social context-dependent *preference function* over outcomes, and *beliefs* concerning the probability that particular actions lead to particular outcomes. Every argument that I have seen for rejecting the rational actor model I have found to be specious, often disingenuous and reflecting badly on the training of its author.

Introduction

Choice behavior can generally be best modeled using the rational actor model, according to which individuals have a time-, state-, and social context-dependent *preference function* over outcomes, and *beliefs* concerning the probability that particular actions lead to particular outcomes. Individuals of course value outcomes besides the material goods and services depicted in economic theory. Moreover, actions may be valued for their own sake. For example, there are *character virtues,* including honesty, loyalty, and trustworthiness, that have intrinsic moral value, in addition to their effect on others or on their own reputation. Moreover, social actors generally value not only *self-regarding* payoffs such as personal income and leisure, but also *other-regarding* payoffs, such as the welfare of others, environmental integrity, fairness, reciprocity, and conformance with social norms (Gintis 2016).

The rational choice model *expresses* but does not *explain* individual preferences. Understanding the content of preferences requires rather deep forays into the psychology of goal-directed and intentional behavior (Haidt 2012), evolutionary theory (Tooby and Cosmides 1992), and problem-solving heuristics (Gigerenzer and Todd 1999). Moreover, the social actor's preference function will generally depend on his current motivational state, his previous experience and future plans, and the social situation that he faces.

H. Gintis (✉)
Santa Fe Institute, Santa Fe, NM, USA
e-mail: hgintis@comcast.net

© Springer International Publishing AG, part of Springer Nature 2018
G. Bronner and F. Di Iorio (eds.), *The Mystery of Rationality*,
https://doi.org/10.1007/978-3-319-94028-1_8

The first principle of rational choice is that in any given situation, which may be time-, state-, and social-context dependent, the decision-maker, say Alice, has a *preference relation* $\succ$ over choices such that Alice prefers x to y if and only if $x \succ y$. The conditions for the existence of such a relation, developed in Section "The Axioms of Rational Choice" below, are quite minimal, the main condition being that Alice's choices must be *transitive* in the sense that if the choice set from which Alice must choose is X with $x, y, z \in X$, then if Alice prefers x to y, and also prefers y to z, then Alice must prefer x to z as well. An additional requirement is that if Alice prefers x to y when the choice set is X, she must continue to prefer x to y in any choice set that includes both x and y. This condition can fail if the choice set itself represents a substantive social context that affects the value Alice places upon x and y. For instance, Alice may prefer fish (x) to steak (y) in a restaurant that also serves lobster (z) because the fish is likely to be very fresh in this case, whereas in a restaurant that does not serve lobster, the fish is likely to be less fresh, so Alice prefers steak (y) to fish (x). For another commonplace example, Alice may prefer a \$100 sweater to a \$200 sweater in a store in which the latter is the highest price sweater in the store, but might reverse her preference were the most expensive sweater in the store priced at \$500. In cases such as these, a more sophisticated representation of choice sets and outcomes both satisfies the rationality assumptions and more insightfully models Alice's social choice situation.

Every argument that I have seen for rejecting the rational actor model I have found to be specious, often disingenuous and reflecting badly on the training of its author. The standard conditions for rationality, for instance, do not imply that rational Alice chooses what is in her best interest or even what gives her pleasure. There are simply *no utilitarian or instrumental implications* of these axioms. If a rational actor values giving to charity, for instance, this does not imply that he gives to charity in order to increase his happiness. A martyr is still a martyr even though the act of martyrdom may be extremely unpleasant. Nor does the analysis assume that Alice is in any sense selfish, calculating, or amoral. Finally, the rationality assumption does not suggest that Alice is "trying" to maximize utility or anything else. The maximization formulation of rational choice behavior, which we develop below, is simply an analytical convenience, akin to the least action principle in classical mechanics, or predicting the behavior of an expert billiards player by solving a set of differential equations. No one believes that light "tries to" minimize travel time, or that billiards players are brilliant differential equation solvers.

The second principle of rational choice applies when Alice's behavior involves *probabilistic* outcomes. Suppose there are a set of alternative possible *states of nature* Ω with elements $\omega_1, \ldots, \omega_n$ that can possibly materialize, and a set of out-comes X. A *lottery* is a mapping that specifies a particular outcome $x \in X$ for each state $\omega \in \Omega$. Let the set of such lotteries be $\mathcal{L}$, so any lottery $\pi \in \mathcal{L}$ gives Alice outcome $x_i = \pi(\omega_i)$ in case ω_i, where $i = 1, \ldots, n$. By our first rationality assumption, Alice has a consistent preference function over the lotteries in $\mathcal{L}$. Adding to this a few rather innocuous assumptions concerning Alice's preferences (see Section "The Axioms of Rational Choice"), it follows that Alice has a consistent preference function $u(\pi)$ over the lotteries in $\mathcal{L}$ and also Alice attaches a specific probability $p(\omega)$ to each event in Ω. This probability distribution is called

Alice's *subjective prior,* or simply her *beliefs,* concerning the events in Ω. Moreover, given the preference function $u(\pi)$ and the subjective prior $p(\omega)$, Alice prefers lottery π to lottery ρ, that is $\pi \succ \rho$, precisely when the *expected utility* of π exceeds that of lottery ρ (see Eq. 8.1).

The rational actor model does not hold universally (see Section "Limitations of the Rational Actor Model"). There are only two substantive assumptions in the above derivation of the expected utility theorem. The first is that Alice does not suffer from *wishful thinking.* That is, the probability that Alice implicitly attaches to a particular outcome by her preference function over lotteries does not depend on how much she stands to gain or lose should that outcome occur. This assumption is certainly not always justified. For instance, believing that she might win the state lottery may give Alice more pleasure while waiting for it to happen than the cost of buying the lottery ticket. Moreover, there may be situations in which Alice will *underinvest* in a desirable outcome unless she inflates the probability that the investment will pay off (Benabou and Tirole 2002). In addition, Alice may be substantively irrational, having excessive confidence that the world conforms to her ideological preconceptions.

The second substantive assumption is that the state of nature that materializes is not affected by Alice's choice of a lottery. When this fails the subjective prior must be interpreted as a *conditional probability,* in terms of which the expected utility theorem remains valid (Stalnaker 1968). This form of the expected utility theorem is developed in Section "The Axioms of Rational Choice".

Of course, an individual may be rational in this decision-theoretic sense, having consistent preferences and not engaging in wishful thinking, and still fail to conform to higher canons of rationality. Alice may, for instance, make foolish choices that thwart her larger objectives and threaten her well-being. She may be poorly equipped to solve challenging optimization problems. Moreover, being rational in the decision-theoretic sense does not imply that Alice's beliefs are in any way reasonable, or that she evaluates new evidence in an insightful manner.

The standard axioms underlying the rational actor model are developed in von Neumann and Morgenstern (1944) and Savage (1954). The plausibility and generality of these axioms are discussed in Section "The Axioms of Rational Choice", where we replace Savage's assumption that beliefs are purely personal "subjective probabilities" with the notion that the individual is embedded in a *network of social actors* over which information and experience concerning the relationship between actions and outcomes is spread. The rational actor thus draws on a network of beliefs and experiences distributed among the social actors to which he is informationally and socially connected. By the sociological principle of *homophily,* social actors are likely to structure their network of personal associates according to principles of social similarity, and to alter personal tastes in the direction of increasing compatibility with networked associates (McPherson et al. 2001; Durrett and Levin 2005; Fischer et al. 2013).

It is important to understand that the rational actor model says *nothing* about how individuals form their subjective priors, or in other words, their *beliefs.* This model does say that whatever their beliefs, new evidence should induce them to transform their beliefs to be more in line with this evidence. Clearly there are many beliefs that are so strong that such updating does not occur. If one believes that something is true with probability one, then no evidence can lead to the Bayesian updating of that belief, although it could lead the individual to revise his whole belief system (Stalnaker 1996). More commonly, strong believers simply discount the uncomfortable evidence. This is no problem for the rational actor model, which simply depicts behavior rather than showing that rational choice leads to objective truth.

The Axioms of Rational Choice

The word *rational* has many meanings in different fields. Critics of the rational actor model almost invariably attach meanings to the term that lie quite outside the bounds of rationality as used in decision theory, and incorrectly reject the theory by referring to these extraneous meanings. We here present a set of axioms, inspired by Savage (1954), that are sufficient to derive the major tools of rational decision theory, the so-called *expected utility theorem.*[1]

A *preference function* $\succeq$ on a choice set Y is a binary relation, where $\{x \succeq y | Y\}$ is interpreted as the decision-maker weakly preferring x to y when the choice set is Y and $x, y \in Y$. By "weakly" we mean that the decision-maker may be indifferent between the two. We assume this binary relation has the following three properties, which must hold for any choice set Y, for all $x, y, z \in Y$, and for any set $Z \subset Y$:

1. *Completeness*: $\{x \succeq y | Y\}$ or $\{y \succeq x | Y\}$;
2. *Transitivity*: $\{x \succeq y | Y\}$ and $\{y \succeq z | Y\}$ imply $\{x \succeq z | Y\}$;
3. *Independence of irrelevant alternatives*: For $x, y \in Z$, $\{x \succeq y | Z\}$ if and only if $\{x \succeq y | Y\}$.

Because of the third property, we need not specify the choice set and can simply write $x \succeq y$. We also make the rationality assumption that the actor chooses his most preferred alternative. Formally, this means that given any choice set A, the individual chooses an element $x \in A$ such that for all $y \in A, x \succeq y$. When $x \succeq y$, we say "x is weakly preferred to y" because the actor can actually be indifferent between x and y.

One can imagine cases where completeness might fail. For instance an individual may find all alternatives so distasteful that he prefers to choose none of them. However, if "prefer not to choose" is an option, it can be added to the choice set with an appropriate outcome. For instance, in the movie *Sophie's Choice,* a woman is asked to choose one of her two children to save from Nazi extermination. The cost of the option "prefer not to choose" in this case was having both children exterminated.

Note that the decision-maker may have absolutely no grounds to choose x over y, given the information he possesses. In this case we have *both* $x \succeq y$ and $y \succeq x$. In this

---

[1] I regret using the term "utility" which suggests incorrectly that the theorem is related to philosophical utilitarianism or that it presupposes that all human motivation is aimed at maximizing pleasure or happiness. The weight of tradition bids us retain the venerable name of the theorem, despite its co notational baggage.

case we say that the individual is *indifferent* between x and y and we write $x \sim y$. This notion of indifference leads to a well-known philosophical problem. If preferences are transitive, then it is easy to see that indifference is also transitive. However it is easy to see that because humans have positive sensory thresholds, indifference cannot be transitive over many iterations. For instance, I may prefer more milk to less in my tea up to a certain point, but I am indifferent to amounts of milk that differ by one molecule. Yet starting with one teaspoon of milk and adding one molecule of milk at a time, eventually I will experience an amount of milk that I prefer to one teaspoon.

The transitivity axiom is implicit in the very notion of rational choice. Nevertheless, it is often asserted that intransitive choice behavior is observed (Grether and Plott 1979; Ariely 2010). In fact, most such observations satisfy transitivity when the state dependence (see Gintis 2007 and Section "Choice Under Uncertainty" below), time dependence (Ahlbrecht and Weber 1995; Ok and Masatlioglu 2003), and/or social context dependence (Brewer and Kramer 1986; Andreoni 1995; Cookson 2000; Carpenter et al. 2005) of preferences are taken into account.

Independence of Irrelevant Alternatives fails when the relative value of two alternatives depends on other elements of the choice set Y, but as suggested above, the axiom can usually be restored by suitably redefining the choice situation (Gintis 2009).

The most general situation in which the Independence of Irrelevant Alternatives fails is when the choice set supplies independent information concerning the *social frame* in which the decision-maker is embedded. This aspect of choice is analyzed in Section "State-Dependent Preferences", where we deal with the fact that preferences are generally state-dependent; when the individual's social or personal situation changes, his preferences will change as well. Unless this factor is taken into account, rational choices may superficially appear inconsistent.

When the preference relation $\succeq$ is complete, transitive, and independent from irrelevant alternatives, we term it *consistent*. It should be clear from the above that preference consistency is an extremely weak condition that is violated only when the decision-maker is quite lacking in reasonable principles of choice.

If $\succeq$ is a consistent preference relation, then there will always exist a utility function such that individuals behave as if maximizing their utility functions over the sets Y from which they are constrained to choose. Formally, we say that a utility function $u : Y \to \mathbf{R}$ *represents* a binary relation $\succeq$ if, for all $x, y \in Y$, $u(x) \geq u(y)$ if and only if $x \succeq y$. We have the following theorem, whose simple proof we leave to the reader.

Theorem 1 *A binary relation $\succeq$ on the finite set Y can be represented by a utility function $u : Y \to \mathbf{R}$ if and only if $\succeq$ is consistent.*

As we have stressed before, the term "utility" here is meant to have no utilitarian connotations.

Choice Under Uncertainty

We now assume that an action determines a *statistical distribution* of possible outcomes rather than a single particular outcome. Let X be a finite set of outcomes and let $\mathcal{A}$ be a finite set of actions. We write the set of pairs (x, a) where x is an outcome and a is an action as $X \times \mathcal{A}$. Let $\succeq$ be a consistent preference relation on $X \times \mathcal{A}$; i.e., the actor values not only the outcome, but the action itself. By Theorem 1 we can associate $\succeq$ with a utility function $u : X \times \mathcal{A} \to \mathbf{R}$.

Let Ω be a finite set of *states of nature*. For instance, Ω could consist of the days of the week, so a particular state $\omega \in \Omega$ can take on the values Monday through Sunday, or Ω could be the set of permutations (about 8×10^{67} in number) of the 52 cards in a deck of cards, so each $\omega \in \Omega$ would be a particular shuffle of the deck. We call any $A \subseteq \Omega$ an *event*. For instance, if Ω is the days of the week, the event "weekend" would equal the set {Saturday, Sunday}, and if Ω is the set of card deck permutations, the event "the top card is a queen" would be the set of permutations (about 6×10^{66} in number) in which the top card is a queen.

Following Savage (1954) we show that if the individual has a preference relation over lotteries (functions that associate states of nature $\omega \in \Omega$ with outcomes $x \in X$) that has some plausible properties, then not only can the individual's preferences be represented by a preference function, but also we can infer the probabilities the individual implicitly places on various events (his so-called *subjective priors*), and the expected utility principle holds for these probabilities.

Let $\mathcal{L}$ be a set of lotteries, where a *lottery* is now a function $\pi : \Omega \to X$ that associates with each state of nature $\omega \in \Omega$ an outcome $\pi(\omega) \in X$. We suppose that the individual chooses among lotteries without knowing the state of nature, after which the state $\omega \in \Omega$ that he obtains is revealed, so that if the individual chooses action $a \in A$ that entails lottery $\pi \in \mathcal{L}$, his outcome is $\pi(\omega)$, which has payoff $u(\pi(\omega), a)$.

Now suppose the individual has a preference relation $\succ$ over $\mathcal{L} \times \mathcal{A}$. That is, the individual values not only the lottery, but the action that leads to a particular lottery. We seek a set of plausible properties of $\succ$ that together allow us to deduce (a) a utility function $u : \mathcal{L} \times \mathcal{A} \to \mathbf{R}$ corresponding to the preference relation $\succ$ over $X \times \mathcal{A}$; and (b) there is a probability distribution $p : \Omega \to \mathbf{R}$ such that the expected utility principle holds with respect to the preference relation $\succ$ over $\mathcal{L}$ and the utility function $u(\cdot, \cdot)$; i.e., if we define

$$\mathbf{E}_\pi[u|a; p] = \sum_{\omega \in \Omega} p(\omega) u(\pi(\omega), a), \tag{8.1}$$

then for any $\pi, \rho \in \mathcal{L}$ and any $a, b \in A$,

$$(\pi, a) \succ (\rho, b) \Leftrightarrow \mathbf{E}_\pi[u|a; p] > \mathbf{E}_\rho[u|b; p]. \tag{8.2}$$

A set of axioms that ensure (8.2), which is called the *expected utility principle,* is formally presented in Gintis (2009). Here I present these axioms more descriptively and omit a few uninteresting mathematical details. The first condition is the rather trivial assumption that

A1. If π and ρ are two lotteries, then whether $(\pi, a) \succ (\rho, b)$ is true or false depends only on states of nature where π and ρ have different outcomes.

This axiom allows us to define a *conditional preference* $\pi \succ_A \rho$, where $A \subseteq \Omega$, which we interpret as "π is strictly preferred to ρ, conditional on event A." We define the conditional preference by revising the lotteries so that they have the same outcomes when $\omega \notin A$. Because of axiom A1, it does not matter what we assign to the lottery outcomes when $\omega \notin A$. This procedure also allows us to define $\succeq_A$ and $\sim_A$ in a similar manner. We say $\pi \succeq_A \rho$ if it is false that $\rho \succ_A \pi$, and we say $\pi \sim_A \rho$ if $\pi \succeq_A \rho$ and $\rho \succeq_A \pi$.

The second condition is equally trivial, and says that a lottery that gives an outcome with probability one is valued the same as the outcome:

A2. If π pays x given event A and action a, and ρ pays y given event A and action b, and if $(x, a) \succ (y, b)$, then $\pi \succ_A \rho$, and conversely.

The third condition asserts that the decision-maker's subjective prior concerning likelihood that an event A occurs is *independent* from the payoff one receives when A occurs. More precisely, let A and B be two events, let (x, a) and (y, a) be two available choices, and suppose $(x, a) \succ (y, a)$. Let π be a lottery that pays x when action a is taken and $\omega \in A$, and pays some z when $\omega \notin A$. Let ρ be a lottery that pays y when action a is taken and $\omega \in B$, and pays z when $\omega \notin B$. We say event A is *more probable than* event B, given x, y, and a if $\pi \succ \rho$. Clearly this criterion does not depend on the choice of z, by A1. We assume a rather strong condition:

A3. If A is more probable than B for some x, y, and a, then A is more probable than B for any other choice of x, y, and a.

This axiom, which might be termed the *no wishful thinking condition,* is often violated when individuals assume that states of nature tend to conform to their ideological preconceptions, and where they reject new information to the contrary rather than update their subjective priors (Risen 2015). Such individuals may have consistent preferences, which is sufficient to model their behavior, but their wishful thinking often entails pathological behavior. For instance, a healthy individual may understand that a certain unapproved medical treatment is a scam, but change his mind when he acquires a disease that has no conventional treatment. Similarly, an individual may attribute his child's autism to an immunization injection and continue to believe this in the face of extensive evidence concerning the safety of the treatment.

The fourth condition is another trivial assumption:

A4. Suppose the decision-maker prefers outcome x to any outcome that results from lottery ρ. Then the decision-maker prefers a lottery π that pays x with probability one to ρ.

We then have the following expected utility theorem:

Theorem 2 *Suppose A1–A4 hold. Then there is a probability function p on the state space Ω and a utility function $u : X \to \mathbf{R}$ such that for any $\pi, \rho \in \mathcal{L}$ and any $a, b \in \mathcal{A}$, $(\pi, a) \succ (\rho, b)$ if and only if $\mathbf{E}_\pi[u|a; p] > \mathbf{E}_\rho[u|b; p]$.*

We call the probability p the individual's *subjective prior* and say that A1–A4 imply *Bayesian rationality,* because they together imply Bayesian probability updating. Because only A3 is problematic, it is plausible to accepted Bayesian rationality except in cases where some form of wishful thinking occurs, although there are other, rather exceptional, circumstances in which the expected utility theorem fails (Machina 1987; Starmer 2000).

Bayesian Updating with Radical Uncertainty

The only problematic axiom among those needed to demonstrate the expected utility principle is the "wishful thinking" axiom A3. While there are doubtless many cases in which at least a substantial minority of social actors engage in wishful thinking, there is considerable evidence that Bayesian updating is a key neural mechanism permitting humans to acquire complex understandings of the world given severely underdetermining data (Steyvers et al. 2006).

For instance, the spectrum of light waves received in the eye depends both on the color spectrum of the object being observed and the way the object is illuminated. Therefore inferring the object's color is severely underdetermined, yet we manage to consider most objects to have constant color even as the background illumination changes. Brainard and Freeman (1997) show that a Bayesian model solves this problem fairly well, given reasonable subjective priors as to the object's color and the effects of the illuminating spectra on the object's surface.

Several students of developmental learning have stressed that children's learning is similar to scientific hypothesis testing (Carey 1985; Gopnik and Meltzoff 1997), but without offering specific suggestions as to the calculation mechanisms involved. Recent studies suggest that these mechanisms include causal Bayesian networks (Glymour 2001; Gopnik and Schultz 2007; Gopnik and Tenenbaum 2007). One schema, known as *constraint-based learning,* uses observed patterns of independence and dependence among a set of observational variables experienced under different conditions to work backward in determining the set of causal structures compatible with the set of observations (Pearl 2000; Spirtes et al. 2001). Eight-month-old babies can calculate elementary conditional independence relations well enough to make accurate predictions (Sobel and Kirkham 2007).

Two-year-olds can combine conditional independence and hands-on information to isolate causes of an effect, and four-year-olds can design purposive interventions to gain relevant information (Glymour et al. 2001; Schultz and Gopnik 2004). "By age four," observe Gopnik and Tenenbaum (2007), "children appear able to combine prior knowledge about hypotheses and new evidence in a Bayesian fashion" (p. 284). Moreover, neuroscientists have begun studying how Bayesian updating is implemented in neural circuitry (Knill and Pouget 2004).

For instance, suppose an individual wishes to evaluate a hypothesis h about the natural world given observed data x and under the constraints of a background repertoire T. The value of h may be measured by the Bayesian formula.

$$P_T(h|x) = \frac{p_T(x|h)P_T(h)}{\sum_{h' \in T} P_T(x|h')P_T(h')}. \tag{8.3}$$

Here, $P_T(h|x)$ is the likelihood of the observed data x, given h and the background theory T, and $P_T(h)$ gives the likelihood of h in the agent's repertoire T. The constitution of T is an area of active research. In language acquisition, it will include predispositions to recognize certain forms as grammatical and not others. In other cases, T might include physical, biological, or even theological heuristics and beliefs.

State-Dependent Preferences

Preferences are obviously state-dependent. For instance, Bob's preference for aspirin may depend on whether or not he has a headache. Similarly, Bob may prefer salad to steak, but having eaten the salad, he may then prefer steak to salad. These state-dependent aspects of preferences render the empirical estimation of preferences somewhat delicate, but they present no theoretical or conceptual problems.

We often observe that an individual makes a variety of distinct choices under what appear to be identical circumstances. For instance, an individual may vary his breakfast choice among several alternatives each morning without any apparent pattern to his choices. Is this a violation of rational behavior? Indeed, it is not.

Following Luce and Suppes (1965) and McFadden (1973), I represent this situation by assuming the individual has a utility function over bundles $x \in X$ of the form.

$$u(x) = v(x) + \epsilon(x) \tag{8.4}$$

where $v(x)$ is a stable underlying utility function and $\epsilon(x)$ is a random error term representing the individual's current idiosyncratic taste for bundle x. This utility function induces a probability distribution π on X such that the probability that the individual chooses x is given by

$$p_x = \pi\{x \in X | \forall\, y \in X, \upsilon(x) + \epsilon(x) > \upsilon(y) + \epsilon(y)\}.$$

We assume $\sum_x p_x = 1$, so the probability that the individual is indifferent between choosing two bundles is zero. Now let $B = \{x \in X | p_x > 0\}$, so B is the set of bundles chosen with positive probability, and suppose B has at least three elements. We can express the Independence of Irrelevant Alternatives in this context by the assumption (Luce 2005) that for all $x, y \in B$,

$$\frac{p_{yx}}{p_{xy}} = \frac{P[y|\{x,y\}]}{P[x|\{x,y\}]} = \frac{p_y}{p_x}.$$

This means that the relative probability of choosing x versus y does not depend on whatever other bundles are in the choice set. Note that $p_{xy} \neq 0$ for $x, y \in B$. We then have

$$p_y = \frac{p_{yz}}{p_{zy}} p_z \tag{8.5}$$

$$p_x = \frac{p_{xz}}{p_{zx}} p_z, \tag{8.6}$$

where $x, y, z \in B$ are distinct, by the Independence of Irrelevant Alternatives. Dividing the first equation by the second in (8.5), and noting that $p_y/p_x = p_{yx}/p_{xy}$, we have

$$\frac{p_{yx}}{p_{xy}} = \frac{p_{yz}/p_{zy}}{p_{xz}/p_{zx}}. \tag{8.7}$$

We can write

$$1 = \sum_{y \in B} p_y = \sum_{y \in B} \frac{p_{yx}}{p_{xy}} p_x,$$

so

$$p_x = \frac{1}{\sum_{y \in B} p_{yx}/p_{xy}} = \frac{p_{xz}/p_{zx}}{\sum_{y \in B} p_{yx}/p_{zy}}, \tag{8.8}$$

where the second equality comes from (8.7).

Let us write

$$\omega(x, z) = \beta \ln \frac{p_{xz}}{p_{zx}},$$

so (8.8) becomes

$$p_{x,B} = \frac{e^{\beta\omega(x,z)}}{\sum_{y\in B} e^{\beta\omega(y,z)}}.$$ (8.9)

But by the Independence of Irrelevant Alternatives, this expression must be independent of our choice of z, so if we write $\omega(x) = \ln p_{xz}$ for an arbitrary $z \in B$, we have

$$p_x = \frac{e^{\beta\omega(x)}}{\sum_{y\in B} e^{\beta\omega(y)}}.$$ (8.10)

Note that there is one free variable, β in Eq. (8.10). This represents the degree to which the individual is relatively indifferent among the alternatives. As $\beta \to \infty$, the individual chooses his most preferred alternative with increasing probability, and with probability one in the limit. As $\beta \to 0$, the individual becomes more indifferent to the alternative choices.

This model helps explain the compatibility of the *preference reversal* phenomenon (Lichtenstein and Slovic 1971; Grether and Plott Grether and Plott 1979; Tversky et al. 1990; Kirby and Herrnstein 1995; Berg et al. 2005) with the rationality postulate. As explained in Gintis (2007), in the cases discussed in the experimental literature, the experimenters offer only alternative lotteries with expected values that are very close to being equal to one another. Thus decision-makers are virtually indifferent among the choices based on the expected return criterion, so even a small influence of the social frame in which the experimenters embed the choice situation on the subjects' preference state may strongly affect their choices. For experimental support for this interpretation, see Sopher and Gigliotti (1993).

Networked Minds and Distributed Cognition

I have stressed that there is one assumption in the derivation of the rational actor model that conflicts with the repeatedly observed fact that human minds are not isolated instruments of ratiocination, but rather are networked and cognition is distributed over this network. I propose here an analytical tool, based on a refinement of the rational actor model proposed by Gilboa and Schmeidler (2001), for representing distributed cognition. Following Gilboa and Schmeidler we assume there is a single decision-maker, say Alice, who faces a *problem p* such that each *action a* that Alice takes leads to some *result r*. Alice does not know the probability distribution of outcomes following action a, so she searches her memory for similar problems she has faced in the past, the action she has taken for each problem, and the result of her action. Thus her memory M consists of a set of *cases* of the form $(q,$

a, r), where q is a problem, a is the action she took facing this problem, and r was the result of the action. Alice has a utility function $u(r)$ defined over results, and a *similarity function* $s(p, q)$ representing how "similar" her current problem p is to any past problem q that she has encountered. Gilboa and Schmeidler then present a set of plausible axioms that imply Alice will choose her action a to maximize the expression.

$$\sum_{(q,a,r)\in M_a} s(p,q)u(r) \qquad (8.11)$$

where M_a is the subset of Alice's memory where she took action a.

Several empirical studies have shown that this case-based decision approach is superior to other more standard approaches to choice under radical uncertainty (Gayer et al. 2007; Golosnoy and Okhrin 2008; Ossadnik et al. 2012; Guilfoos and Pape 2016). To extend this to distributed cognition, we simply replace Alice's personal memory bank by a wider selection of cases distributed over her social network of minds. It would also be plausible to add a second similarity function indicating how similar the individual who actually took the action is Alice herself.

Limitations of the Rational Actor Model

One often hears that a theory fails if there is a single counterexample. Indeed, this notion was the touchstone of Karl Popper's famous interpretation of the scientific method (Popper 2002[1959]). Because biological systems are inherently complex, this criterion is too strong for the behavioral sciences (Godfrey-Smith 2006, 2009; Weisberg 2007; Wimsatt 2007). Despite its general usefulness, the rational actor model fails to explain choice behavior in several well-known situations. Two examples are the famous Allais and Ellsberg Paradoxes. These are of course not paradoxes, but rather violations of rational choice.

The Allais Paradox

Allais (1953) offered the following scenario as a violation of rational choice behavior. There are two choice situations in a game with prizes $x = \$2,500,000$, $y = \$500,000$, and $z = \$0$. The first is a choice between lotteries $\pi = y$ and $\pi' = 0.1x + 0.89y + 0.01z$. The second is a choice between $\rho = 0.11y + 0.89z$ and $\rho' = 0.1x + 0.9z$. Most people, when faced with these two choice situations, choose $\pi \succ \pi'$ and $\rho \succ \rho'$. Which would you choose?

This pair of choices is not consistent with the expected utility principle. To see this, let us write $u_h = u(2500000)$, $u_m = u(500000)$, and $u_l = u(0)$. Then if the expected utility principle holds, $\pi \succ \pi'$ implies $u_m > 0.1u_h + 0.89u_m + 0.01u_l$, so

$0.11u_m > 0.10u_h + 0.01u_l$, which implies (adding $0.89u_l$ to both sides) $0.11u_m + 0.89u_l > 0.10u_h + 0.9u_l$, which says $\rho \succ \rho'$.

Why do people make this mistake? Perhaps because of *regret*, which does not mesh well with the expected utility principle (Loomes 1988; Sugden 1993). If you choose π' in the first case and you end up getting nothing, you will feel really foolish, whereas in the second case you are probably going to get nothing anyway (not your fault), so increasing the chances of getting nothing a tiny bit (0.01) gives you a good chance (0.10) of winning the really big prize. Or perhaps because of *loss aversion*, because in the first case, the anchor point (the most likely outcome) is $500,000, while in the second case the anchor is $0. Loss-averse individuals then shun π', which gives a positive probability of loss whereas in the second case, neither lottery involves a loss, from the standpoint of the most likely outcome.

The Allais paradox is an excellent illustration of problems that can arise when a lottery is consciously chosen by an act of will and one *knows* that one has made such a choice. The regret in the first case arises because if one chose the risky lottery and the payoff was zero, one knows for certain that one made a poor choice, at least ex post. In the second case, if one received a zero payoff, the odds are that it had nothing to do with one's choice. Hence, there is no regret in the second case.

But in the real world, most of the lotteries we experience are chosen by default, not by acts of will. Thus, if the outcome of such a lottery is poor, we feel bad because of the poor outcome but not because we made a poor choice.

The Ellsberg Paradox

Another classic violation of the expected utility principle was suggested by Ellsberg (1961). Consider two urns. Urn A has 51 red balls and 49 white balls. Urn B also has 100 red and white balls, but the fraction of red balls is unknown. One ball is chosen from each urn but remains hidden from sight. Subjects are asked to choose in two situations. First, a subject can choose the ball from urn A or urn B, and if the ball is red, the subject wins $10. In the second situation, the subject can choose the ball from urn A or urn B, and if the ball is white, the subject wins $10. Many subjects choose the ball from urn A in both cases. This violates the expected utility principle no matter what probability the subject places on the probability p that the ball from urn B is white. For in the first situation, the payoff from choosing urn A is $0.51u(10) + 0.49u(0)$ and the payoff from choosing urn B is $(1 - p)u(10) + pu(0)$, so strictly preferring urn A means $p > 0.49$. In the second situation, the payoff from choosing urn A is $0.49u(10) + 0.51u(0)$ and the payoff from choosing urn B is $pu(10) + (1 - p)u(0)$, so strictly preferring urn A means $p < 0.49$. This shows that the expected utility principle does not hold.

Whereas the other proposed anomalies of classical decision theory can be interpreted as the failure of linearity in probabilities, regret, loss aversion, and epistemological ambiguities, the Ellsberg paradox appears to strike even more deeply because it implies that humans systematically violate the following principle of first-order stochastic dominance (FoSD).

Let $p(x)$ and $q(x)$ be the probabilities of winning x or more in lotteries A and B, respectively. If $p(x) > q(x)$ for all x, then $A \succ B$.

The usual explanation of this behavior is that the subject *knows* the probabilities associated with the first urn, while the probabilities associated with the second urn are *unknown,* and hence there appears to be an added degree of risk associated with choosing from the second urn rather than the first. If decision-makers are risk-averse and if they perceive that the second urn is considerably riskier than the first, they will prefer the first urn. Of course, with some relatively sophisticated probability theory, we are assured that there is in fact no such additional risk, so it is hardly a failure of rationality for subjects to come to the opposite conclusion. The Ellsberg paradox is thus a case of performance error on the part of subjects rather than a failure of rationality.

Failures of Judgment

Contemporary behavioral economics has developed a powerful critique of the standard assumption that people are instrumentally rational (Ariely 2010; Thaler and Sunstein 2008). In fact, human decision-makers are close to instrumentally rational when they are sufficiently informed and the cost of exploring alternative strategies is low (Gintis 2009; Gigerenzer 2015). Nevertheless, the behavioral economics critique of the assumption of instrumental rationality is important and well-taken.

But as we have seen, the rational actor model depicts *formal rationality,* not *instrumental rationality.* That is, it assumes that people have consistent preferences and update according to Bayes rule, but it does not assume that rational behavior is oriented towards any particular end state or goal, and certainly not that rational behavior furthers the fitness or welfare interests of the decision-maker. Let us review the major claims made by behavioral economists supporting the notion that choice behavior is fundamentally irrational.

- *Logical Fallibility*: Even the most intelligent decision-makers are prone to commit elementary errors in logical reasoning. For example, in one well-known experiment performed by Tversky and Kahneman (1983), a young woman, Linda, is described as politically active in college and highly intelligent. The subject is then asked the relative likelihood of several descriptions of Linda, including the following two: "Linda is a bank teller" and "Linda is a bank teller and is active in the feminist movement." Many subjects rate the latter statement more likely than the former, despite the fact that the most elementary reasoning shows that if p implies q, then p cannot be more likely than q. Because the latter statement implies the former, it cannot be more likely than the former.

- *Anchoring*: When facing extreme uncertainty in making an empirical judgment, people often condition their behavior on recent but irrelevant experience. For instance, suppose a subject is asked to write down a number equal to the last two digits of his social security number and then to consider whether he would pay this number of dollars for particular items of unknown value. If he is then asked to bid for these items, he is likely to bid more if the number he wrote down was higher.

- *Cognitive Bias*: If you ask someone to estimate the result of multiplying $1 \times 2 \times 3 \times 4 \times 5 \times 6 \times 7 \times 8$, he is likely to offer a lower estimate than if you had presented him with $8 \times 7 \times 6 \times 5 \times 4 \times 3 \times 2 \times 1$. Similarly, if you ask someone what fraction of English words end in "ng" and give the example "gong," you will probably get a lower estimate than if you gave the example "going."

- *Availability Heuristic*: People tend to predict the frequency of an event based on how often they have heard about it. For example, most people believe that homicides occur with more frequency than suicides, although the reverse is the case. Similarly, they believe that certain cancers cluster in certain communities because of environmental pollutants, where in fact such clusters may occur no more frequently than chance, but are more likely to be reported.

- *Status Quo Bias*: Decision-makers tend to follow a certain traditional pattern of behaviors even after there is strong credible evidence that a superior course of action is available. For instance, in an early well-known experiment, Samuelson and Zeckhauser (1988) presented subjects with a task in which several financial assets were listed and the subjects were asked to choose one that they prefer to invest in. A second set of subjects was given the same list of financial assets, but one was presented as the *status quo*. They found that the asset listed as the *status quo* was chosen at a much higher frequency than when it was presented just as one among several randomly presented alternatives.

- *Herd Mentality*: People are heavily influenced by the actions of others. For instance, Solomon Asch (1951) showed that peer pressure can induce subjects to offer clearly false evaluations, even when the subject and his peers do not know each other and will likely never meet outside the laboratory. Groups were formed consisting of eight college students, all but one of whom were confederates of the experimenter. Each student was shown a card with a black line on it, and a second card with three black lines, one of which was the same length as the line on the first card, and the other two were of very different lengths, one longer and the other shorter. Each student was asked to say out loud which line on the second card matched the line on the first card, the seven confederates going first and choosing an obviously incorrect line. More than a third of subjects agreed with the obviously wrong answer.

- *Framing Effects*: A framing effect is a form of cognitive bias that occurs when choice behavior depends on the wording of two logically equivalent statements. Take, for example, the classic example of the physician and his heart patient, analyzed by McKenzie et al. (2006) and Thaler and Sunstein (2008). The patient must decide whether to have heart surgery or not. His doctor tells him either (A) "Five years after surgery, 90% of patients are alive," or (B) "Five years after surgery, 10% of patients are dead." The two statements are of course logically equivalent, but subjects are far more likely to accept surgery with frame (A) than with frame (B).

- *Default Effects*: In choosing among various options, if one is offered as the default option, people tend to choose it with high frequency. A most dramatic example is organ donation (Johnson and Goldstein 2003). Countries in Europe that have a presumed consent default have organ donation rates that are about 60% higher than countries with explicit consent requirements. Another famous example involves registering new employees in a company 401(k) savings plan. When participating is the default, participation is considerably higher than when the default is non-participation (Bernheim et al. 2011).

The logical fallibility argument would of course be devastating to rational choice theory, which implicitly assumes that decision-makers are capable of making logical deductions. There are certainly complex logical arguments that the untrained subject is likely to get wrong. Indeed, even a mistake in mathematical computation counts as an error in logical reasoning. But what appear to be the elementary errors of the type revealed by the Linda the Bank Teller example are more likely to be errors of interpretation on the part of the experimenters. It is important to note that given the description of Linda, the probability that an individual is Linda if we know that the individual is a bank teller is much lower than the probability that an individual is Linda if we know that she is a feminist bank teller. This is because Linda is probably a feminist, and there are far fewer feminist bank tellers than there are bank tellers. Subjects in the experiment might reasonably assume that the experimenters were looking for a conditional probability response rather than a simple probability response because they supplied a mass of information that is relevant to conditional probability, but is quite irrelevant to simple probability.

Indeed, in normal human discourse, a listener assumes that any information provided by the speaker is relevant to the speaker's message (Grice 1975). Applied to this case, the norms of discourse reasonably lead the subject to believe that the experimenter wants Linda's politically active past to be taken adequately into account (Hilton 1995; Wetherick 1995). Moreover, the meaning of such terms as "more likely" or "higher probability" are vigorously disputed even in the theoretical literature, and hence are likely to have a different meaning for the average subject versus for the expert. For example, if I were given two piles of identity folders and asked to search through them to find the one belonging to Linda, and one of the piles was "all bank tellers" while the other was "all bank tellers who are active in the feminist movement," I would surely look through the latter (doubtless much smaller) pile first, even though I am well aware that there is a "higher probability" that Linda's folder is in the former pile rather than the latter one. In other words, conditional rather than straight probability is the appropriate concept in this case.

However important anchoring, cognitive bias, and the availability heuristic may be, they are clearly not in conflict with the rational actor model because they do not compromise any of the rational choice axioms. In particular, they do not imply preference inconsistency or the failure of Bayesian updating. The *status quo* bias may seem to contradict Bayesian updating, but it does not. For one thing, if one is satisfied with a particular choice, it may plausibly appear excessively costly to evaluate properly new information. Herbert Simon (1972) called this reasonable behavior "satisficing." It is clearly compatible with rational updating. For another, one may reasonably ignore new information on the grounds that it is unreliable. Models of Bayesian updating simply assume that the new information is rigorously factual, which is often not the case.

The framing effects literature is more challenging. Indeed, some argue that because it is impossible to avoid framing effects, there are no true underlying preferences, so the rational actor model fails. This conclusion is unwarranted. In this book we specified from the outset that preferences are generally state-, time-, and social frame-dependent. In particular, preferences are frame-dependent because individual choices, except perhaps for Robinson Crusoe before he meets Friday, occur within a social context, and that context is the social frame for choice behavior. Indeed, even the absence of a social frame is a social frame.

Consider, for instance, the physician and his heart patient scenario described above. Thaler and Sunstein (2008) interpret this as showing that many decision-makers are irrational. But it is more accurate to interpret these results as patients simply following the implicit suggestion of the physician, the expert on whom their well-being depends. Note first that neither (A) nor (B) gives the patient sufficient information to make an informed choice because the physician does not provide the equivalent survival and death rates *without* surgery. The only reasonable inference is that the patient believes the doctor is recommending surgery in case (A), and recommending against surgery in case (B).

The physician and his heart patient example is not an isolated case of the tendency for behavioral economists to ignore the intimately social nature of choice, and to interpret completely reasonable behavior as irrational. Gigerenzer (2015), who documents several additional examples of this tendency, concludes:

> Research… indicates that logical equivalence is a poor general norm for understanding human rationality… Speakers rely on framing in order to implicitly convey relevant information and make recommendations, and listeners pay attention to these. In these situations, framing effects clearly do not demonstrate that people are mindless, passive decision-makers.

Similarly, default effects do not illustrate the decision-maker's irrationality, but rather the tendency to treat the default as a recommendation by experts whose advice it is prudent to follow unless there is good information that the default is not the best choice (Johnson and Goldstein 2003). Indeed, Gigerenzer (2015) reports that a systematic review of hundreds of framing studies could not find a single one showing that framing effects incur real costs in terms.

References

Ahlbrecht, M., & Weber, M. (1995). Hyperbolic discounting models in prescriptive theory of intertemporal choice. *Zeitschrift fur Wirtschafts- und Sozialwissenschaften, 115,* 535–568.

Allais, M. (1953). Le comportement de l'homme rationnel devantle risque, critique des postulats et axiomes de l'ecole Americaine. *Econometrica, 21,* 503–546.

Andreoni, J. (1995). Warm-glow versus cold-prickle: The effects of positive and negative framing on cooperation in experiments. *Quarterly Journal of Economics, 110*(1), 1–21.

Ariely, D., & Irrational, P. (2010). *The hidden forces that shape our decisions.* New York: Harper.

Asch, S. (1951). Effects of group pressure on the modification and distortion of judgments. In H. Guetzkow (Ed.), *Groups, leadership and men,* (pp. 177–190). Carnegie Press.

Benabou, R., & Tirole, J. (2002). Self confidence and personal motivation. *Quarterly Journal of Economics, 117*(3), 871–915.

Berg, J. E., Dickhaut, J. W., & Rietz, T. A. (2005). *Preference reversals: The impact of truth-revealing incentives.* Lowa: College of Business, University of Lowa.

Bernheim, B. D., Fradkin, A., & Popov, I. (2011). The welfare economics of default options in 401 (k) plans. *American Economic Review, 105*(9), 2798–2837. (National Bureau of Economic Research, No. w17587.).

Brainard, D. H., & Freeman, W. T. (1997). Bayesian color constancy. *Journal of the Optical Society of America A, 14,* 1393–1411.

Brewer, M. B., & Kramer, R. M. (1986). Choice behavior in social dilemmas: effects of social identity, group size, and decision framing. *Journal of Personality and Social Psychology, 50* (543), 543–549.

Carey, S. (1985). *Conceptual change in childhood.* Cambridge: MIT Press.

Carpenter, J. P., Burks, S. V., & Verhoogen, E. (2005). Comparing student workers: The effects of social framing on behavior in distribution games. *Research in Experimental Economics, 1,* 261–290.

Cookson, R. (2000). Framing effects in public goods experiments. *Experimental Economics, 3,* 55–79.

Durrett, R., & Levin, S. A. (2005). Can stable social groups be maintained by Homophilous imitation alone? *Journal of Economic Behavior & Organization, 57*(3), 267–286.

Ellsberg, D. (1961). Risk, ambiguity, and the savage axioms. *Quarterly Journal of Economics, 75,* 643–649.

Fischer, I., Frid, A., Goerg, S. J., Levin, S. A., Rubenstein, D. I., & Selten, R. (2013). Fusing enacted and expected mimicry generates a winning strategy that promotes the evolution of cooperation. In *Proceedings of the National Academy of Sciences, 110*(25), 10229–10233.

Gayer, G., Gilboa, Itzhak, & Lieberman, O. (2007). Rule-based and case-based reasoning in housing prices. *The BE Journal of Theoretical Economics, 7,* 1.

Gigerenzer, G. (2015). On the supposed evidence for Libertarian Paternalism. *Re of Philosophical Psychology, 6,* 361–383.

Gigerenzer, G., & Todd, P. M. (1999). *Simple heuristics that make us smart.* New York: Oxford University Press.

Gilboa, I., & Schmeidler, D. (2001). *A theory of case-based decisions.* Cambridge: Cambridge University Press.

Gintis, H. (2007). A framework for the unification of the behavioral sciences. *Behavioral and Brain Sciences, 30*(1), 1–61.

Gintis, H. (2009). *The bounds of reason: Game theory and the unification of the behavioral sciences.* Princeton: Princeton University Press.

Gintis, H. (2016). *Individuality and entanglement: The moral and material bases of human social life.* Princeton: Princeton University Press.

Glymour, A., Sobel, D. M., Schultz, L., & Glymour, C. (2001). Causal learning mechanism in very young children: Two- three- and four-year-olds infer causal relations from patterns of variation and covariation. *Developmental Psychology, 37*(50), 620–629.

Glymour, C. (2001). *The mind's arrows: Bayes nets and graphical causal models in psychology.* Cambridge: MIT Press.

Godfrey-Smith, P. (2006). The strategy of model-based science. *Biology and Philosophy, 21,* 725–740.

Godfrey-Smith, P. (2009). Models and fictions in science. *Philosophical Studies, 143,* 101–116.

Golosnoy, V., & Okhrin, Y. (2008). General uncertainty in portfolio selection: A case-based decision approach. *Journal of Economic Behavior & Organization, 67*(3), 718–734.

Gopnik, A., & Meltzoff, A. (1997). *Words, thoughts, and theories.* Cambridge: MIT Press.

Gopnik, A., & Tenenbaum, J. B. (2007). Bayesian networks, bayesian learning and cognitive development. *Developmental Studies, 10*(3), 281–287.

Gopnik, A., & Schultz, L. (2007). *Causal learning, psychology, philosophy, and computation.* Oxford: Oxford University Press.

Grether, D., & Plott, C. R. (1979). Economic theory of choice and the preference reversal phenomenon. *American Economic Review, 69*(4), 623–638.

Grice, H. P. (1975). Logic and Conversation. In Donald Davidson & Gilbert Harman (Eds.), *The Logic of Grammar* (pp. 64–75). Encino, CA: Dickenson.

Guilfoos, T., & Pape, A. D. (2016). Predicting human cooperation in the prisoner's Dilemma using case-based decision theory. *Theory and Decision, 80,* 1–32.

Haidt, J. (2012). *The righteous mind: Why good people are divided by politics and religion.* New York: Pantheon.

Hilton, D. J. (1995). The social context of reasoning: Conversational inference and rational judgment. *Psychological Bulletin, 118*(2), 248–271.

Johnson, E. J., & Goldstein, D. G. (2003). Do defaults save lives? *Science, 302,* 1338–1339.

Kirby, K. N., & Herrnstein, R. J. (1995). Preference reversals due to myopic discounting of delayed reward. *Psychological Science, 6*(2), 83–89.

Knill, D., & Pouget, A. (2004). The Bayesian brain: The role of uncertainty in neural coding and computation. *Trends in Cognitive Psychology, 27*(12), 712–719.

Lichtenstein, S., & Slovic, Paul. (1971). Reversals of preferences between bids and choices in gambling decisions. *Journal of Experimental Psychology, 89,* 46–55.

Loomes, G. (1988). When actions speak louder than prospects. *American Economic Review, 78*(3), 463–470.

Luce, R. D. (2005). *Individual choice behavi or.* New York: Dover.

Luce, R. D., & Suppes, P. (1965). Preference, utility, and subjective probability. In R. D. Luce, R. R. Bush, & E. Galanter (Eds.), *Handbook of mathematical psychology,* (Vol. III). New York: Wiley.

Machina, M. J. (1987). Choice under uncertainty: Problems solved and unsolved. *Journal of Economic Perspectives, 1*(1), 121–154.

McFadden, D. (1973). Conditional logit analysis of qualitative choice behavior. In P. Zarembka (Ed.), *Frontiers in econometrics,* (pp. 105–142). New York: Academic Press.

McKenzie, C. R. M., Liersch, M. J., & Finkelstein, S. R. (2006). Recommendations implicit in policy defaults. *Psychological Science, 17,* 414–420.

McPherson, M., Smith-Lovin, L., & Cook, J. (2001). Birds of a feather: Homophily in social networks. *Annual Review of Sociology, 27,* 415–444.

Ok, E. A., & Masatlioglu, Y. (2003). *A general theory of time preference.* New York: Economics Department, New York University.

Ossadnik, W., Wilmsmann, D., & Niemann, B. (2012). Experimental evidence on case-based decision theory. *Theory and decision, 75*(2), 211–232.

Pearl, J. (2000). *Causality.* New York: Oxford University Press.

Popper, K. (2002[1959]) *The Logic of Scientific Discovery.* London: Routledge Classics.

Risen, J. L. (October 2015). Believing what we do not Believe. *Psychological Review,* 1–27.

Samuelson, W., & Zeckhauser, R. (1988). Status Quo Bias in decision making. *Journal of Risk and Uncertainty, 1,* 7–59.

Savage, L. J. (1954). *The foundations of statistics.* New York: Wiley.

Schultz, L., & Gopnik, A. (2004). Causal learning across domains. *Developmental Psychology, 40,* 162–176.

Simon, H. (1972). Theories of bounded rationality. In C. B. McGuire, & R. Radner (Eds.), *Decision and Organization,* (pp. 161–176). New York: American Elsevier.

Sobel, D. M., & Kirkham, N. Z. (2007). Bayes nets and babies: Infants' developing statistical reasoning abilities and their representations of causal knowledge. *Developmental Science, 10* (3), 298–306.

Sopher, B., & Gigliotti, G. (1993). Intransitive cycles: Rational choice or random error: an answer based on estimation of error rates with experimental data. *Theory and Decision, 35,* 311–336.

Spirtes, P., Glymour, C., & Scheines, R. (2001). *Causation, Prediction, and Search.* Cambridge: MIT Press.

Stalnaker, R. (1968). A theory of conditionals. In Nicholas Rescher (Ed.), *Studies in Logical Theory.* London: Blackwell.

Stalnaker, R. (1996). Knowledge, belief, and counterfactual reasoning in games. *Economics and Philosophy, 12,* 133–163.

Starmer, C. (2000). developments in non-expected utility theory: The hunt for a descriptive theory of choice under risk. *Journal of Economic Literature, 38,* 332–382.

Steyvers, M., Griffiths, T. L., & Dennis, S. (2006). Probabilistic inference in human semantic memory. *Trends in Cognitive Sciences, 10,* 327–334.

Sugden, R. (1993). An Axiomatic Foundation for Regret Theory. *Journal of Economic Theory, 60* (1), 159–180.

Thaler, R. H., & Sunstein, Cass. (2008). *Nudge: Improving Decisions about Health, Wealth, and Happiness.* New York: Penguin.

Tooby, J., & Cosmides, L. (1992). The Psychological Foundations of Culture. In J. H. Barkow, L. Cosmides, & J. Tooby (Eds.), *The Adapted Mind: Evolutionary Psychology and the Generation of Culture* (pp. 19–136). New York: Oxford University Press.

Tversky, A., & Kahneman, D. (1983). Extensional versus intuitive reasoning: The conjunction fallacy in probability judgement. *Psychological Review, 90,* 293–315.

Tversky, A., Slovic, P., & Kahneman, D. (1990). The Causes of Preference Reversal. *American Economic Review, 80*(1), 204–217.

von Neumann, J., & Oskar, M. (1944). *Theory of Games and Economic Behavior.* Princeton: Princeton University Press.

Weisberg, M. (2007). Who is a Modeler? *British Journal of the Philosophy of Science, 58,* 207–233.

Wetherick, N. E. (1995). Reasoning and rationality: A critique of some experimental paradigms. *Theory & Psychology, 5*(3), 429–448.

Wimsatt, W. C. (2007). *Re-Engineering Philosophyfor Limited Beings.* Cambridge: Harvard University Press.

Chapter 9
Rationality and Irrationality Revisited or Intellectualism Vindicated or How Stands the Problem of the Rationality of Magic?

Ian Jarvie

Abstract The problem of rationality emerges in anthropology when actions and ideas strike the ethnographer as inconsistent with commonsense or with science (e.g. to get crops you plant seeds; chanting is neither here nor there). The commonsense and the science appealed to are those of the society to which the anthropologist belongs or into which he has been socialized. Hence the problem is intrinsically ethnocentric. Rationality assessments are comparative judgments with no presupposition that perfect rationality is anywhere achieved or achievable. Rationality is normative: a work in progress. We strive to be ever more consistent whilst knowing we can never achieve perfection. An alternative proposal appears in a recent paper by Lukes on rationality and in Sørensen's critique of my intellectualist view. These are criticized in their turn and it is reaffirmed that judgments of comparative rationality do little or no explanatory work. Beliefs and assertions as such are neither rational nor irrational; it is only our actions regarding them that can be so assessed. Explanation proceeds by matching means to postulated ends in defined situations, whether the ends are articulated or must be supplied by the social scientist as hypotheses to be tested. Ritual actions like chanting are pervasive. The place of ritual in science and the place of ritual in magic are quite different.

The problem of the rationality of magic is simple enough. One set of actions, such as planting some seeds and doing a chant, presuppose practical and scientific means-end rationality (planting seeds to get crops) as well as, at the same time, supernatural intervention (hence chanting to get crops). Planting without chanting would seem unproblematically rational. Chanting without planting would seem unproblematically irrational. It is the combination that creates the problem. How can the actors both be rational and irrational regarding the one end (crops)? Does the chant make the whole set of actions irrational, that is, lacking means-ends rationality?

I. Jarvie (✉)
York University, Toronto, Canada
e-mail: jarvie@yorku.ca

© Springer International Publishing AG, part of Springer Nature 2018
G. Bronner and F. Di Iorio (eds.), *The Mystery of Rationality*,
https://doi.org/10.1007/978-3-319-94028-1_9

An underlying assumption is that manipulating a cause to achieve an effect is unproblematically rational. If we want to explain the chanting we reconstruct what the actors believe about it, what they think it does. This is known as "intellectualism", and following Durkheim, some anthropologists disagree with it along the following lines. Magical chanting is not undertaken because it is thought to bring about the result (crops). Chanting does not cause crops to grow and the chanters are quite aware of this. Something else altogether is going on. Chanting is (not causal but) symbolic, it is about creating and asserting meanings around whatever is chanted over. Chanting over the planted seeds is about investing the agricultural with a sense of its social importance, warranting a collective ceremony. These are not implausible claims. Take agricultural practice in North America. It would not be hard to find communities where before or after preparing the field and planting, the farmers and their families (perhaps a majority of them) gather in their Christian churches and say prayers that plead for an abundant harvest. The farmers would likely be "scientific" farmers, fully aware of how to adapt their field preparation to local conditions, to disease threats, and to the vagaries of the weather. They cannot be saddled with the idea that their prayer meetings are part of their scientific farming. Even if we complicate the example with superstitions we will not find this mistake. If a superstitious farmer always starts from a good luck corner of the same field, or ties a red ribbon to the lead tractor, he does not think the start point or the ribbon initiate some part of the causal chain that leads to crops. If there is drought in the land and the whole community, week after week, prays for timely rain, they do not see the prayer as part of the scientific text book on crops (or on the weather).

Hence, the anthropologists who disagree with intellectualism conclude, it is wholly incorrect to explain magical, superstitious, or religious ceremonies by alleged background beliefs in their causal efficacy. They are not part of the causal nexus, they are part of a different, symbolic or meaningful, nexus of human activity, the nexus that brings humans together to engage in collective gestures of solidarity and social significance. Social cohesion, then, is the true end of the chanting and of the prayer.

On this anthropological view there are two realms simultaneously present: the nexus of cause and effect, acknowledged by practical common sense and by scientific investigation and wholly naturalistic; and the nexus of ritual, symbols, and meanings, which are social rather than natural, which are not about cause and effect, and which function to serve the community needs in a different way from the natural nutritional service offered by the crops. Such an analysis of the natural and the social, with the supernatural confined to its functions for the latter, might be called the Durkheimian riposte to Tylor and Frazer, the original so-called "intellectualist" theorists of magic. The claim that the two realms coexist serves to block the intellectualist claim that the magician, perhaps also the religious, base their practices on mistakes about what causes what. Traditional anthropological subject peoples may seldom be as scientific farmers as those of North America but they are not under false impressions about cause and effect. They do not think their magic is an alternative to irrigation. In fact, they will learn to irrigate and continue their magical rituals.

Whether the manipulation of symbols and meanings for social benefit is rational or not thus becomes moot. If rationality and judgments of rationality are confined to the nexus of cause and effect then it would seem that any appeal to the socially symbolic and meaningful could at best be *arational*, neither rational nor irrational. It is of course always open to the inquirer to broaden the conception of rationality and to thus allow that ritual actions that serve some social function are rational too. Any such functional rationality is of a different order from the rationality of matching cause to desired effect, since functionalists usually deny teleology. They deny, that is, that the ritual actors are aware that they are promoting social ends. Whether such a double rationality is a worthwhile intellectual structure is a matter for another day.

There is here a complication seldom faced: the refusal to see the chanting or the prayer as a part of the agricultural act rests on the assumption that the farmer is aware of the indifference of the farming to the chanting or the prayer; yet the assumption that the function of the act is a part of the contribution of the individual to social cohesion disregards the farmer's ignorance of this function. Either we assume that awareness is essential to action or that it is irrelevant to it; why assume that it matters here but not there? For, all actions have a symbolic component, and if we declare this component the cause of inefficient action even with no awareness, then the question arises as to the reason for it when already the efficient action is symbolic. To put it more circumspectly, all actions, including the most practical, can be interpreted to serve both their ostensible purposes of bringing about aimed-at ends and other socially desirable states as well. And then the purely ritual part of the act is redundant, and so the Durkheimian explanation of it is no explanation at all.

Take an example from science: scientists gathering around the tea and coffee machine may be both swapping ideas and thus promoting intellectual exchange *and* reinforcing social networks that improve the morale of the lab. In fact the two aspects of the interactions may in practice be indistinguishable. Again, the Large Hadron Collider at CERN may have detected the Higgs boson and tested assumptions on its specific properties, as was hoped and intended, and it also is (and was understood to be) a heartening symbol of international cooperation. By parallel reasoning, the planting itself, regardless of the chanting, has symbolic meaning. It is usually a cooperative activity, even if sometimes a small-scale one, and can serve all the functions of collective activity. Indeed, it is a collective representation *par excellence*.

None of this is surprising. Science is *par excellence* a social rather than a solitary activity, and the actions of scientists taken within their social institutions will have social consequences. The consequences just mentioned, networking and international cooperation, are intended. But it is a truth universally acknowledged by social scientists that there can be no action and interaction without unintended social consequences, some desirable, some less so. The functionalists present a rosy picture of the positive social benefits of the chanting and the praying. They are less alert to the ways in which the symbolic might by dysfunctional, might even impede the naturalistic nexus of cause and effect. If the symbolic can interfere with the accomplishment of practical ends then resort to it is irrational.

Obviously, if the chanting does not leave time or energy for the planting then the symbolic is disruptive of the causal and so not in a separate realm. The potlatch comes to mind. It is a feast, so nutrition is offered and so are gifts of value. But the striking issue around the potlatch has always been the claim that it could get out of hand and become socially destructive. A bit like the old chestnut of the tribe that perished for its reverence for nature, especially for man-eating tigers. Less far-fetched is some of the disorder and destruction sewn by Melanesian cargo cults.

I have already given an intellectualist account of these cults. Their being treated as irrational was troubling.[1] There was a logic to them as well as a destructiveness.[2] The kind of intellectualism I embraced is known as "critical rationalism". In this paper I look at some alternatives of Lukes and Horton, Wittgenstein, and Sørensen, as foils to formulate forcefully the thesis of critical rationalism that judgments of rationality are of limited explanatory value to the social sciences. Rational action, I contend, is a matter of context or situation and always comes in degrees, and rationality is a property neither of belief nor of believing. Rationality, it turns out, comes down to "being explicable". The only cases of zero rationality are those we call "inexplicable", "miraculous", "sheer luck", etc.

Cargo cults are religious phenomena of twentieth century Melanesia. They are syncretic cults that combine elements of traditional mythology and world view (such as the ancestors waiting to return) with redemption stories that smack of diffused Christian teaching. A prophet arises and teaches that the world is either ending or is going to be turned upside down. Prominent among the promises are that cargo, or material goods, will be delivered to the native peoples and not to the storehouses of the merchants and colonial authorities. Some cults prophesy a political upheaval whence the oppressed and exploited will become powerful and their present rulers powerless. Some even postulate that there will be an exchange of skin colour. Williams in the 1920s was influential in labelling them irrational because he found their ideas disordered and because scarce resources of labour and capital were squandered on arrangements to receive the prophecy and its upshot.[3]

What does it amount to characterize these cargo cult actions as irrational? How does labelling them 'irrational' advance social science? How does the charge of irrationality function in our cognitive economy? The claim that an action is irra-tional does not explain it. Rather the reverse: to say that an act is irrational is often to admit that it is baffling or inexplicable, especially when the actors in question are in other respects reasonable or competent individuals.

In the case of cargo cults the explanation I offered reconstructed the logic of the cultists' situation, in line with the methodology that recommends social

---

[1] As when Williams called them a "madness", see F. E. Williams and J. H. P. Murray, *The Vaillala Madness and the Destruction of Native Ceremonies in the Gulf Division, Papua: Anthropology Reports No. 4*, Port Moresby: AMS Press.

[2] *The Revolution in Anthropology*, London: Routledge and Kegan Paul, 1964. Lawrence (1965).

[3] F. E. Williams and J. H. P. Murray, *The Vaillala Madness and the Destruction of Native Ceremonies in the Gulf Division, Papua: Anthropology Reports No. 4*, Port Moresby: AMS Press. Williams lamented that cargo cults disrupted traditional native life.

explanations be analyzed according to situational logic. A situational reconstruction consists of explanatory premises that describe actors' ends along with their situation, namely their social and material conditions as well as their grasp of these, from which, by the application of the rationality principle, once can deduce a description of the expected outcome. Such an approach was called "intellectualist" because the definition of the situation included a reconstruction of the cultists' views and values.

In a 1967 article I wrote with Agassi ("The Problem of the Rationality of Magic") we made the general claim that rationality always comes in degrees (people can act more or less adequately to their situation), and that rationality can be assessed only in relation to the situation as defined. Judgment by the social scientist, however ethnocentric, was inescapable and could not be replaced by an algorithm. Relativizing judgments of rationality to the situation is not the same as affirming multiple rationalities in each and every cultural context. Allowing multiple rationalities struck us as a recipe for total breakdown of the project of explanatory social science.

Our work received a boost when Bryan Wilson included it in his anthology, *Rationality* (1970).[4] That volume included papers by Steven Lukes and Martin Hollis who offered different barriers to multiple rationalities: coherence and logic as universal values. It is generally agreed that coherence and logic are a necessary condition for judging rationality of thought systems; we contested the claim that they are sufficient. When they are said to suffice, human fallibility is precluded. For our part, we thought of rationality more as a project, as fallible attempts to extend and apply standards of coherence and logic. Different levels of achievement make for different situational logics. Furthermore, we had a deeper difference with Hollis and Lukes: we regarded logic as the organon of **criticism**, not of **justification**. Criticism we thought of in a positive way, as constituting in itself a contribution to knowledge rather than merely as destructive of obstacles to its acquisition.

Lukes returned to the issue in his chapter "The Problem of Apparently Irrational Beliefs" (Turner and Risjord, *Philosophy of Anthropology and Sociology*,[5] 2007). Lukes distinguishes two main positions on rationality in anthropology: the rationalist (or universalist) approach—to which he was formerly wedded—and the localist (or pluralist or relativist) approach that advocates multiple rationalities with their own standards. He characterizes himself in 2007 as 'in search of some coherent combination and reconciliation of the two approaches that retains what is plausible in each' (Lukes 2007, 600). Unfortunately for his project, Lukes seeks guidance in the later work of the social anthropologist Robin Horton. Horton was once an advocate of fallibilism and openness to criticism, something of an honorary critical rationalist. Lukes looks instead to Horton's later work, which turned justificationist and naturalistic.

Horton splits human cognitive interaction with the environment to the Primary World and the Secondary World. The Primary World comprises objects that are

---

[4]Wilson (1970).

[5]Turner and Risjord (2007).

directly experienced, such as mountains, rivers, sticks, stones, animals, and people. Interaction with this Primary World is governed by Primary Theory that comprises instrumental reason, basic logic, and observation. It is common to all human groups, involving as it does the primary processes of nutrition, nurture, shelter, reproduction, etc. Its presence can be mapped across cultures and across time with a large degree of mutual intelligibility. It is the common denominator of our common existence. Let us call it the Brass Tacks theory.[6]

Beyond Brass Tacks is the Secondary World and Secondary Theory. The objects inhabiting **it** are largely hidden and thus open to free speculation, thus able to display a startling variety across cultures and time. Unobservable entities and powers are not checkable by common experience and basic reasoning. So they vary greatly between cultures, at times even within them. Horton adds that he considers these two realms "primary" and "secondary" because of the order in which they develop in the individual, in society, and most likely in humanity as well. Lukes suggests that Horton has made concrete what philosophers have sometimes spoken of as points of contact, as bridgeheads, between different systems of thought. Horton's proposal he finds concrete, reasonable, even charitable. What, then, does Lukes consider irrational? Answer: belief and hence action in the Secondary World using Secondary Theory and reasoning, insofar as it violates or goes beyond Primary World rules of reasoning and delves into the unseen. Secondary Theory allows loose reasoning which protects irrational tales of mysteries and miracles.

Lukes has produced an ingenious new formulation of the philosophy of the Enlightenment, one that presents science as the outcome of employing a rigorous refinement and development of the rules of Primary Theory that are universal while strictly avoiding the Secondary that is local and thus prone to error and irrationality.

Admittedly, this view is deeply appealing to an Enlightenment Fundamentalist (to use Gellner's term) such as myself. Would that I could endorse it. Yet I am wary: the self-same deliverance was also promised by the logical positivist verification theory of meaning (and by Hume before that). The anti-meaninglessness project has vanished almost without trace. On the Horton-Lukes view, all human groups, as long as they focus on the world of 'medium sized dry goods', can reason in sound ways and conduct the business of life in a rationally intelligible manner, but as soon as they strive for explanations that go beyond Brass Tacks they start postulating occult qualities, hidden entities, forces, and processes, using "reasoning" that differs markedly from society to society and from time to time, perhaps even from person to person. Although both Horton and Lukes are aware that science too postulates hidden entities, forces, and processes, they claim that what it uses is but a legitimate extension of Primary Theory and reasoning. The question then is, what differentiates a legitimate extension of theory and reasoning from the Primary World to the Secondary World of illegitimate Secondary Theory? Neither

---

[6]I am irresistibly reminded of T. S. Eliot's *Sweeney Agonistes*: "Birth, and copulation, and death/ That's all the facts when you come to brass tacks". Horton's list is longer and Malinowski's Basic Needs are somewhat similar. The Brass Tacks theory is not the list of brass tacks (see next note) but the idea that all humans calculate logically and coherently when it comes to them.

Horton nor Lukes offers clear criteria for legitimate extensions of Primary Theory to the Secondary World (nor can they). Instead, they point to the fruitfulness of science, i.e., its having empirical character. Such a move cannot but strike one as special pleading, considering that the entities that modern scientific theory postulates are more mysterious than those of magic theory. Their exemption for science is thus question-begging; hence it is illegitimate by their own naturalistic lights. Horton and Lukes owe themselves a clear account of how Primary Theory and Primary Reason can legitimately be extended to the hidden world, and to state what extensions are not legitimate. The naturalistic demarcation between the legitimate and the illegitimate that that their theory calls for must draw a sharp line between science and magic. Since science grew out of magic and religion,[7] this cannot be done. In any case, it takes little common sense to agree that at times, no matter how rarely, magic succeeds and science fails. These rare cases, alas, suffice to refute the naturalism of Horton and Lukes.

My alternative to their naturalism is a conventional demarcation that captures at least the difference between crude magic (chanting) and the best science (say, Nobel-Prize quality work). Even the best demarcation proposal will be open to criticism that, if successful, will prompt efforts at improvement. To allow for this procedure all we need is to replace efforts to justify with efforts to improve. A prominent criterion of demarcation is empirical falsifiability. It directs us to sharpen magical and religious claims, to draw from them empirical consequences which can be and at times are tested; the procedure embodies the scientific attitude and can be used to explain the gradual transition from magic to science as due to gradual refinement of the rules of reasoning, i.e. institutional innovation. Whether this rational reconstruction of mine fits the historical facts is a moot question; not moot is that magic preceded science and still coexists alongside it despite repeated scientistic exorcism.

Following Horton, Lukes distinguishes the basic reasoning around Brass Tacks as logic and observation. What is different in the reasoning around the Secondary World? Lukes does not say. Nor can he. Previously, anthropologists have suggested that thinking about the Secondary World employs reasoning by association, analogy, metaphor, symbol, narrative. Some philosophers of science have shown interest in these forms of reasoning (if such they be). My view is that these labels are simply colorful characterizations of the premises used in deductive argument and that reasoning from them is universal and deductive or it is faulty. If there is any whiff of induction then the reasoning, whether associational, analogical, metaphorical, symbolic, or narrative, is invalid. The invalidity can be pinpointed, but only with speculative and criticizable premises, not justified ones.

By the view of Lukes and Horton that much Secondary Theory is irrational, we swim in a sea of irrationality, since layers of magical and superstitious theory coexist with our contemporary scientific outlook. By declaring so much to be irrational we create unnecessary problems for the social sciences. What aim is

---

[7]Popper (1963).

served by characterizing much common belief and action as irrational? "Irrational" is condemnatory, it is a form of judgment with an overtone of disapproval. If everyday life is awash in irrationality what now should we call the irrationality of those in mental hospitals? Sometimes we can make their outlook intelligible for their case files. But they are so much at odds with the rest of society that at times their presence within it warrants quarantine. Yet insanity has little or nothing to do with magic and superstition. Lumping everyday beliefs and action together with those of the mentally ill is crude from the perspective of social science. It raises a very aggressive question: does the view of Horton and Lukes apply equally to the superstitious and to the mentally ill? Should we recommend treating them all the same way?

Not that I advocate unlimited charity to the magically-minded and the superstitious. Lukes gives short shrift to the idea that magic and superstition are expressive and/or symbolic, a proposal blatantly made in order to excuse magic and superstition from being ranked on the scale of rational/irrational altogether and thus to take away from their practitioners the ability to improve. The primary symbolist error is that it is posed as an alternative to intellectualism, sometimes it is even postulated as a distinct realm of human endeavour, of making meaning. Yet it is open to us to read any action or utterance, no matter how instrumental, as having symbolic significance. The symbolic is always a potential and never the whole story. Hence it has little or no explanatory value except to those with a religious and essentialist tendency, i.e. holders of suppressed premises about the make-up of the world. The symbolist view is very popular with charity-minded anthropologists and philosophers, since it shields the subject thought systems from critical scrutiny. Yet this unlimited charity is unlimited condescension. It was forcefully expressed in some notes Wittgenstein wrote on his copy of Frazer's *The Golden Bough* that were published posthumously.[8] He expressed the view that there were no underlying magical thought systems to be critically assessed, that, as he put it, magical rituals, expressive actions, and symbols, do not rest on opinions and so they cannot be judged as errors. Viewed this way they are not open to improvement; yet how else then did science emerge from this nexus? Perhaps Wittgenstein would say that it did not.

Wittgenstein was an engineer, a logician, and a philosopher of language; not surprisingly, he advocated excessive rationalism; it drove him to mysticism and irrationalism. He did not put his own irrationalism in the same category as theories whose verisimilitude can be assessed. Indeed, he denounces the project of seeking to explain magico-religious phenomena, yet he explained religion away as a mere feeling of being immune to all danger. He asserted that apart from animal-like actions, some ritual actions are best approached thus: "We can only *describe* and say: human life is like that." (*RFGB* 3e). This is part-and-parcel of his contempt for the making of explanatory hypotheses: "There must not be anything hypothetical in

---

[8]Wittgenstein was deeply contemptuous of the English: "How narrow is the spiritual life for Frazer. Thus, how impossible to understand another life in terms of the English life of his time! Frazer can imagine no priest who is not basically an English parson of our time, with all his stupidity and dullness" (Tambiah 60). Wittgenstein's views do nothing to curb his ethnocentrism.

our considerations. We must do away with all explanation and description alone must take its place" (*Philosophical Investigations*, I, 109).

Wittgenstein offered the following, argument: "When I am angry about something, I sometimes hit with my stick on the ground or a tree, etc. But certainly I don't believe that it is the fault of the ground or that hitting can help. 'I can release my anger.' And all rites are of this kind. Such actions one can call actions of instinct." (Tambiah 56).[9] Anthropologists have a name for such reasoning, they call it the "if I were a horse" argument. It amounts to saying, "this is what I would do in that situation therefore that is what those people are doing".

Regrettably, Wittgenstein here equated error with stupidity, and failed to allow that there can be error that is in no way stupid, and that still possesses expressive and symbolic dimensions. Since magical and religious ideas have been held to be the origin of certain scientific ideas, deeming them error is far from calling them stupid. They may contain error, but that is a deficiency, not a stupidity. (To put it Wittgenstein's way, within a form of life we do not call what belongs to it "stupidity".) This all seems rather obvious. Where do Horton and Lukes stand on this issue? On this question their theory stands or falls.

The association of stupidity with error is commonplace, and casts a pall on much history of science, where the simple conjecture and refutation model is sometimes evaded precisely in order to avoid saying plainly and unequivocally that this or that scientific conjecture turned out to be false. Would it not be more effective to defend the magician by arguing that he does not think about his magical rituals in any systematic way, any more than we think systematically about all our rituals of, for example, the way we greet one another? This line of defence would raise problems too: just because the underlying thought-system is not articulated, it does not follow that it is not there. This is an important point which I shall press.

It is a standard task of the ethnographer to reconstruct the shards and patches, hints and glimpses of a magical thought system. The difficulties it involves are candidly set out by Evans-Pritchard in his *Witchcraft, Oracles, and Magic among the Azande* (1937). To deny the value of such reconstruction, or to insist that magic is all instinct, or expressive symbolism, is a philistine assault on the very project of ethnographic science. *Contra* Wittgenstein, Popper has proposed the guiding rule that "we are not to abandon the search for universal laws and for a coherent theoretical system, nor ever give up our attempts to explain causally any kind of event we can describe" (Popper 1959, §12). This is a proposed methodological rule, not a *diktat*. I know of no convincing argument against it. Wittgenstein offered none.

To be credible, Horton and Lukes should add a clearer-cut demarcation between Brass Tacks theory and Secondary World theory. Instead, they naturalistically point at the tangibility of the former and contrast it with the hiddenness of the latter. This is not good enough. Since Newton, all theories of science postulate a world of hidden, invisible, forces and entities, even hidden in principle. Some historians of

---

[9]Tambiah (1990).

science take the view that a lot of science takes its cue from the hidden by assuming it and then working out empirical consequences from it and testing them. It is this systematic drive to pin things down, make them concrete, and link them to the observable world, that marks the scientific attitude. The short label for it is the demand for falsifiability. The demand is a proposed rule, a convention, or a social institution, and it is to be adopted or rejected with arguments about its fruitfulness: does it do a better job of coordinating our address to the world than any rival rule?

Having argued that there is a magico-religious world-view behind magic and religion, and that this system of ideas deserves to be criticized and tested wherever possible, I turn now to the notion of belief itself and its rational status.

Underneath the seeming irrationality of many actions is simple inconsistency, failure to confront conflicts of evidence. To the actors each of their actions seems reasonable. To the social scientist that reasonableness disappears as soon as one step back is taken and the action is viewed in a wider context. Each context provides the possibility of rational reassessment. Realization that context-shifting can change the assessment of rationality immediately leads to the discovery that context should be the first consideration, "rational in what context?" the first question. Magic is rational enough in the context of the magico-religious world-view or context. It is less rational in the context of the scientific world view.[10] The scientific context is a rational improvement on the magical context, since it can account for both. Moreover, unlike the religious world view, and, to a lesser extent, the magical world view, the scientific world view does not require belief, notwithstanding what most philosophers of science say. Science as such is indifferent to your state of belief. Scientists can cooperate without checking each other's beliefs. (Scientists do not ask if their peers are religious or not; when a scientist advocates an outdated theory, whether for religious reasons or not, their advocacy is ignored, unless they argue rationally for the revival of the dead theory.) Science is a social process, a cooperative process of reasoning and testing, and its upshot is not a belief state but the claim that further action will move us along the road of the growth of knowledge.

Most philosophers take it for granted that the justification of a scientific thesis is universal and so unqualified. They then face the absurdity that refuted classical scientific theories are used in some situations and unrefuted contemporary theories are used in others. Scientists and philosophers of science who notice this fact are sometimes tempted to escape to relativism. Horton and Lukes cannot possibly follow this route. Its difficulty becomes manageable if we view the universalism of science as a preference rather than as an achievement and apply to the best of science the same situational logic that Agassi and I advocate applying to the most primitive magic—on the understanding that rationality is not black-and-white but mostly grey: it comes in degrees. For this we need to eschew the idea that belief can

---

[10]One interesting implication: magic in a magico-religious society is more rational than magic in our society. Along one dimension, then, a simpler yet non-scientific society may be more rational than a scientific society pervaded by magic, as ours is.

be rational plain and simple in science or elsewhere. Here are some arguments for leaving beliefs out of the discussion of rationality.

First argument. Beliefs are rather vaguely specified entities. How do I know you believe something? By a speech act, some say. When do you affirm it by a speech act? When are you serious; not when are you acting. How about when you are self-deceived? How about when you do not understand what you are saying? Mental states like these should be subject to some sort of Occam's razor: if we think we can get by without discussing them then let us try to do so and see how successful we are.

Second argument, rational belief is a relic of the idea that beliefs are optional and reached by calculation. The calculation is a part of the theory of science as inductive that is sham. Options and choices are also unreal. Most beliefs we have are socialized into us, under the calculating radar so to speak. Not that we cannot revise or change them; but not with ease and hardly the whole set at the same time, most of which we are scarcely aware we take for granted. The background only becomes the foreground when something disturbs it, and the more so the more critical it permits us to be.

Third argument, beliefs in the unseen are often irrefutable, hence they are empirically undecidable. Should undecidable beliefs be eschewed? It would seem an excessive demand, for scientific reasons. Undecidable intimations have stimulated many a scientific quest.

Fourth argument, belief is private, and we have a right to that. Freedom to believe is also freedom not to disclose what you believe. We are supposed to tolerate people of different beliefs, not to impose on them those beliefs we hold to be scientific. How does that toleration fit with ranking beliefs on a rationality scale? Would the "science is rational belief" philosophers want us to show up magic and religion in school? The implicit model must not be religious orthodoxy.

A final, more positive argument. The relation between scientific work and belief needs to take account of scientific scepticism and lack of commitment, as so often articulated by Einstein. A line of thought and experiment in science can be pursued entirely in the spirit of "let us see where it leads", and all questions of whether or not anyone believes anything about the matter pushed firmly aside.

These are a few reasons for my reaffirmation of our claim that beliefs are in themselves arational.[11] When we conjecture that ideas of efficacy underlie magic it is a logical reconstruction. When someone starts fixing a car it is logical in reconstructing their actions to include the claim "this car can be fixed" without entering into their mental state or whether they **really** think or believe anything like it. We need to broach questions of rationality only when we want to explain. What we want to explain are apparent irrationalities. These we explain by the situation: the world view, the available information, and the willingness to reflect and to test. Some contexts encourage this more than others. Closed contexts, where the world view is more or less total, however tacit, are less rational than mixed contexts, or,

---

[11]Agassi (1981).

I almost wrote, open contexts. But "open" would be a mistake. All contexts are to be judged open to this or that degree. No society is totally closed, not even Japan before the Meiji restoration. Similarly, no society is completely open. Openness is always, like rationality, a project, an ongoing attempt at self-improvement. Indeed, openness and rationality are closely related.

Since I have thus far contrasted my proposal of how to explain magic both with the alternative proposal of Lukes and with Wittgenstein's objection to all explanation, it seems that there is no other option to discuss. But there is. One critic of my view of magic, Jesper Sørensen, *A Cognitive Theory of Magic* (2007) is not opposed to explanation, not even to explanation of matters social, but he objects specifically to intellectualist explanations of magic. This objection rests on his particular view of magic. He too asserts that magic is not undergirded by a thought system at all and hence cannot be used to explain magical actions. He makes a strong connection between magic and ritual. He says that ritual does not result from thought-systems. "Rather, the performance of ritual actions invoking superhuman agents is a prerequisite for explanations of non-ritual events (such as disease) as caused by such agents" (p. 6). This seems to say that if you explain falling ill as due to witchcraft you can only do so if you perform rituals that involve witchcraft. It is a variation on the nineteenth century view that rituals come first, explanations of them later ("study the ritual not the belief"[12]).

Sørensen's agenda is to craft an evolutionary account of religion and magic. This is not hard to do: I offer him one for free: magic and religion are false science and so evolutionary dead ends. My account will not attract Sørensen, who developed his arguments in a theological context (and presumably, although it is not stated, for a theological purpose. Explanation requires that one reconstruct and fill in). My albeit crude view is that some rituals, such as repetitions in collective work, are explained by the end (planting seeds). Why not explain the associated chanting this way too? If the ritual is part of a package deal then the same end explains both planting and chanting and the enthymeme is completed by reconstructing the missing premises. Reconstruction is necessary because most beliefs are far from fully developed and systematized—or even fully expressed. The shards and patches nature of popular thought-systems is not confined to the tribes anthropologists used to study, but is also the case for most societies. Our own commonsense, much appealed to by philosophers, is hardly a fully articulated, coherent and consistent system of thought. It is full of magic and superstition, including some masquerading as religion. When we invoke it to explain, however, we treat it as articulated, otherwise is has no probative value. Evans-Pritchard in his work on both the Azande and the Nuer freely acknowledges that he builds a picture of witchcraft or of Nuer religion that is much more neat and systematic than the reality on the ground. It is, one might say, a rational reconstruction of their thought systems. The purpose of the

---

[12]For a discussion of this phrase and its role in functionalist anthropology see the subject index entry 'study the ritual, not the belief' in my *The Revolution in Anthropology*, London: Routledge and Kegan Paul 1964.

rational reconstruction is to enable explanation. If social scientists refrain from these exercises, all thought systems become *ad hoc* and hence their explanation is *ad hoc*.

Sørenson's alternative is to avoid the *ad hoc* by considering only magical systems that have a full-fledged ritual apparatus. But then the social scientist needs a theory of magical systems that do not abide by this condition. Also, for those that do the social scientist still needs a theory of ritual. Sørensen criticizes the intellectualist view of magic as false science, he says, because intellectualists do not have a theory of ritual. And why is a theory of ritual needed? Because it is of the essence of matters in the field of the sacred[13] that they are ritualized, he says. Is this generalization true? Rain dances are rituals, but are love potions and "bless you" and "Open Sesame"? Suppose that ritual is inescapable; does it demand of us that we theorize about it? What kind of a criticism is it of a theory to say that it does not contain a theory of an adjacent item? We were discussing not ritual but magic. And even if all magic is ritual, not all ritual is magic. And of course, each adjacent item has an item adjacent to it. Hence, what Sørensen demands of us is not a theory of magic or of anything specific; he demands of us a whole thought system. Let me spell it out: much of magic, religion, and science is not a system but bits and pieces, semi-coherent at best. Some of the greatest achievements of science are due to efforts to achieve coherence and science has never reached full coherence. Even mathematics has not reached it. Much of the magic that fills the modern so-called scientific world is non-ritualized; much of the science is ritualized (but that does not make its concern the sacred). Hence the intellectualist answer would be that our theory of magic as false science can be conjoined with whatever theory of ritual our readers happen to favour, provided it is not inconsistent with the intellectualist outlook. Certainly my own view of ritual is that it is a coordinating and mnemonic institution that has its origins in confusions about cause and effect. (All this is particularly true of rituals that were consciously devised by founders and reformers.)

To sum up. The intellectualist view of magic as false science is entirely compatible with the historical facts of magic, in the form of alchemy and astrology, and of religion as the provider of a single law giver and a single set of natural laws. No demarcation is necessary between **the ideas** that stand behind science, magic and religion, and no special kind of reasoning is allowed to any of them. Rather each of them applies different social conventions and attitudes to their attempts to describe and to explain the world. Their conventions and attitudes can be assessed by how well they help us to bump our conjectures against the world and thus help delineate its outlines. It is intellectual systems that enable us to explain the world, to deduce, that is, consequences that match or fail to match the way the world is. Tidying up, reconstructing, and making articulate are carried out on all our thought, scientific

---

[13]Perhaps his theological background shows in his invocation of the category of the sacred as though it were unproblematic. My own inclination would be to see the sacred simply as a label attached to certain sorts of rituals. Not the scientists round the water cooler ritual, but the scientists on their knees in church.

and non-scientific in the course of assessing our scientific progress. In natural science the exercise is called the literature review and is rightly considered creative.

Consider the Nobel Prize ceremony. The prize winners in physics, medicine, and the like, are scientists. Their work has been conducted in the usual scientific institutions, and has been assessed and reassessed by ever higher level institutions all the way up to the Nobel Committee. Usually the prize winners have discovered something. A paradigm case would be the prize to Wilkins, Watson, and Crick for the double-helix structure of DNA. Once they were awarded the prize an elaborate ritual was initiated. On a given date the recipients were hosted in Stockholm by the Nobel Committee, they gave lectures if so inclined, they were wined and dined, and finally, in evening dress at a formal ceremony, they were presented with their prizes by the King. There were speeches, music, choreography, and rounds of applause. There was a large formal banquet. None of these rituals made their work sacred or to do with the sacred, but it is a flagrant example of how there are symbolic aspects to science. The rituals around receipt of the Nobel Prize are not incantations to produce more and better prize-worthy science, they are celebrations to stress the weight and importance the prize-winning work and to signal that the state of Sweden encourages and rewards the work of science in some sense on behalf of humanity as a whole. Of course, if the Nobel Prize is viewed as a Holy Grail then it could also provide a cause and effect incentive to do science.

What goes on here is a simple fact: writers on magic ignore rituals in science as inessential, on the tacit assumption that in magic ritual is essential. The effort to avoid ethnocentrism is laudable, but it cannot obliterate the fact that in some societies rituals are of the essence and in others not. Nobody will declare the value of a contribution to scientific knowledge or technology less valid if the Nobel Ceremony was wrongly performed. Both the water cooler and the Nobel Ceremony are contingent for scientific achievement; in their absence other rituals would take their place. The symbolic is always with us. Sometimes it is innocuous; sometimes it is helpful; sometimes it thwarts the nexus of cause and effect. The intellectualist view takes ideas seriously and subjects them to scrutiny. Too often claims that there are no ideas behind actions at all, or that symbolizing is being mistaken for ideation, are defensive attempts to shield ideas from scrutiny. Such defensiveness inhibits intellectual progress, for intellectual progress is always made by discovering our mistakes.

Acknowledgements This paper had its origins in a workshop on "Individual and Collective Rationality" held under the auspices of the Institut des Sciences Humaines et Sociales of the University of Nice-Sophia Antipolis on March 21–22 2011. My thanks to the organizers, Richard Arena, Alban Bouvier, and Bernard Conein, for the invitation and the prompt it gave me to revisit these questions.

References

Agassi, J. (1981). "Faith has nothing to do with rationality", reprinted in his *Science and Society*, Dordrecht: D. Reidel.

Lawrence, P. (1965). *Road belong cargo*. Manchester, UK: Manchester University Press.

Lukes, S. (2007). "The problem of apparently irrational beliefs". In P. T. Stephen & W. R. Mark (Eds.), *Philosophy of anthropology and sociology, volume in the handbook of the philosophy of science series* (pp. 591-606). Amsterdam: Elsevier.

Popper, K. R. (1963). "Back to the Pre-Socratics", in his *Conjectures and Refutations*, London: Routledge & Kegan Paul.

Tambiah, S. J. (1990). *Magic, science, religion and the scope of rationality*. Cambridge: Cambridge University Press.

Turner, S. P., & Risjord, M. W. (Eds.). (2007). *Philosophy of anthropology and sociology, handbook of philosophy of sciences*, series. Amsterdam: Elsevier.

Wilson, B. R. (Ed.). (1970). *Rationality*. Oxford, UK: Basil Blackwell.

Chapter 10
Rational Life Plans?

Daniel Little

Abstract What is involved in creating a rational plan for one's life? This article considers the idea of a "rational life plan" and argues that we should not consider a rational life plan as a blueprint for carrying out one's purposes and values. Instead, we should think of the task of living one's life rationally and purposefully as more akin to the project of navigation in unknown terrain, where a few overriding intentions about the intended direction of one's travel are mixed with timely adjustments to obstacles and opportunities. Rationality in living one's life involves thoughtful reflection and deliberation about one's enduring values and purposes, along with incorporation of course corrections as circumstances change.

Introduction

Millions of words have been written on the topic of rationality in action. Life involves choices. How should we choose between available alternatives? Where should I go to college? Which job should I accept? Should I buy a house or rent an apartment? How much time should I give my job in preference to my family? We would like to have reasons for choosing A over B, and we would like to approach these choices "rationally".

These are all "one-off" choices, and rational choice theory has something like a formula to offer for the decider: gain the best knowledge available about the several courses of action; evaluate the costs, risks, and rewards of each alternative; and choose that alternative that produces the greatest expected level of satisfaction of your preferences. There are nuances to be decided, of course: should we go for "greatest expected utility" or should we protect against unlikely but terrible outcomes by using a maximin rule for deciding? John Von Neumann and Oscar Morgenstern formalized these premises of decision theory in *Theory of Games and Economic Behavior* (Von Neumann and Morgenstern 1944).

D. Little (✉)
University of Michigan-Dearborn, Dearborn, USA
e-mail: delittle@umich.edu

© Springer International Publishing AG, part of Springer Nature 2018   131
G. Bronner and F. Di Iorio (eds.), *The Mystery of Rationality*,
https://doi.org/10.1007/978-3-319-94028-1_10

There are several deficiencies in this story. Most obviously, few of us actually go through the kinds of calculations specified here. We often act out of habit or semi-articulated rules of thumb. Moreover, we are often concerned about factors that do not fit into the "preferences and beliefs" framework, like moral commitments, conceptions of ourselves, loyalties to others, and the like. Pragmatists would add that much mundane action flows from a combination of habit and creativity rather than formal calculation of costs and benefits. Finally, researchers like Kahneman and Tversky have demonstrated that human reasoners systematically deviate from these assumptions in fairly predictable ways (Kahneman et al. 1982), and Herbert Simon offered an alternative conception of "bounded rationality" (Simon 1979) that was intended to serve as a better conception of practical decision making across a range of circumstances.

But my concern here is larger. What is involved in being deliberative and purposive about the whole of one's life? How do we lay out the guideposts of a life plan? What is involved in acting deliberatively and purposively in creating and carrying out one's life plan or other very long-term projects? Is there such a thing as a "rational plan of life"?

Here I want to look more closely than usual at what is involved in reflecting on one's purposes and values, formulating a plan for the medium or long term, and acting in the short term in ways that further the big plan. My topic is "rationality in action", but I want to pay attention to the issues associated with large, extended purposes—not bounded decisions like buying a house, making a financial investment, or choosing a college to attend. I am thinking of larger subjects for deliberation involving long periods of time and multiple goods and values.

The special features of a plan for life include several important characteristics: large temporal scope; the malleability of one's fundamental goals, aspirations, and character; substantial uncertainties about the future; and extensive intertwining of moral and political values with more immediate concerns of self-interest, prudence, and desire. Moreover, the act of formulating plans on this scale and living them out is formative: we become different persons through these efforts. So our preferences and desires are not exogenous; rather, they are a partially endogenous result of a series of past deliberations and choices. Part of what we plan for is what kind of person we want to be; and this means undertaking actions and habits that change our values and characteristics as decision-makers.

What role does rational deliberation play in action horizons at this level of scope? To what extent and through what processes can individuals engage in a rational process in thinking through their decisions and plans at this level? Is it an expectation of rationality that an individual should have composed integrated sets of plans and objectives, from the most global to the intermediate to the local? Or instead, does a person's journey through large events take its shape in a more stochastic way: opportunities, short-term decisions, chance involvements, and some ongoing efforts to make sense of it all in the form of a developing narrative? It is possible to maintain that life is not planned, but rather built like Neurath's raft with materials at hand; and that rationality and deliberation come in only at a more local scale.

What Is Involved in the Idea of a "Plan of Life"?

Aristotle, Kant, and Rawls agree on the idea people ought to have rational plans of life to guide their everyday efforts and activities. But what is involved in being rational about one's plan of life? More fundamentally, what is a plan of life? Is it a sketch of a lifetime goal, along with some indications of the efforts that are currently thought to lead to this goal? Is it a blueprint for organizing one's thinking, actions, investments, time, resources, and character over time in order to bring about the intended goal? Or is it something more flexible that this? Did Walter White in *Breaking Bad* have a plan of life, either before and after his cancer diagnosis? Did Dostoevsky have a plan of life? How about Wagner or Whitman? Is it possible to be rational in making partial or full life plans?

It is clear that the concept of planning is crucial to this question. Planning means orchestrating one's activities over time in such a way as to bring about good outcomes over the full period. When a person plans for a renovation of his or her home, he or she considers the reasons for considering the renovation; the results to be achieved; the enhancements that would contribute to those results; the resources that are necessary to fund those enhancements; the amount of time that will be required for each of the sub-tasks; and so forth. With a good plan and a good execution, it is likely that a good outcome will be achieved: an improved residence that was accomplished within the budgeted time and resources available.[1] Planning makes perfectly good sense for the normal bounded projects within one's ordinary life. But does the idea of rational planning extend to the fullness of one's own life?

Perhaps Aristotle is the ancient philosopher who had the greatest interest in the idea of rational life-planning. Aristotle made it clear in the *Nichomachean Ethics* (1987) that he believed that thinking rationally about life and happiness is the most important thing we can devote ourselves to, and it involves thinking clearly about the goods we want to bring about. To be rational about living means to be reflective about the goods that one pursues over an extended period of time. However, John Rawls was the philosopher in recent times who brought this idea into serious attention. The concept plays an important role within his theory of justice in *A Theory of Justice*. Rawls introduces the idea in the context of his discussion of primary goods (Rawls 1971).

> The main idea is that a person's good is determined by what is for him the most rational long-term plan of life given reasonably favorable circumstances. A man is happy when he is more or less successful in the way of carrying out this plan. To put it briefly, the good is the satisfaction of rational desire. We are to suppose, then, that each individual has a rational plan of life drawn up subject to the conditions that confront him. This plan is designed to permit the harmonious satisfaction of his interests. It schedules activities so that various desires can be fulfilled without interferences. It is arrived at by rejecting other plans that are either less likely to succeed or do not provide for such an inclusive attainment of

---

[1]The most interesting contributions to the topic of planning are to be found in the work of David Bratman, including his *Intention, Plans and Practical Reason* (Bratman 1987). Also helpful is Jonathan Baron's *Rationality and Intelligence* (Baron 1985).

aims. Given the alternatives available, a rational plan is one which cannot be improved upon; there is no other plan which, taking everything into account, would be preferable. (Rawls 1971: 92–93)

Several things are noteworthy about this description. First, Rawls specifies that a life plan involves scheduling activities so as to "harmoniously satisfy interests", which is paraphrased as "fulfilling desires without interferences". In other words, Rawls's account of a plan of life is a fairly shallow one in terms of the assumptions it makes about the person. It takes desires as fixed and then "plans" around them to ensure their optimal satisfaction. But there are other things that we might want to include in a plan of life—choices about one's enduring character, for example. And second, Rawls makes a very heroic assumption here by requiring that a rational plan of life is a uniquely best plan, an optimal plan, one that cannot be improved upon. But we now understand more clearly how utterly impossible it is to imagine full optimization for even relatively simple choice problems, let alone the whole of one's life. Herbert Simon's concept of satisficing is plainly more relevant to the current context. So Rawls's conception of a rational plan of life will not be satisfactory. It is too comprehensive, and it takes too many features of life as being fixed in advance (including one's own character and commitments).

Special Features of Life Plans

We may begin by noting that a life plan is not like the more limited choices one makes during one's life. Consider the space of choices that confronts the 20-year-old college student Miguel: what kind of work will satisfy me over the long term? How much importance will I attribute to higher income in twenty years? Do I want to have a spouse and children? How much time do I want to devote to family? Do I want to live in a city or the countryside? How important to me is integrity and consistency with my own values over time? These kinds of questions are difficult to answer in part because they do not yet have answers. Miguel will *become* a person with a set of important values and commitments; but right now he is somewhat plastic. It is possible for him to change his preferences, tastes, values, and concerns over time. So perhaps his plan needs to take these kinds of interventions into account.

Another source of uncertainty has to do with the future of the world itself. Will the economy continue to provide decent opportunities for young people, or will income stratification continue to increase? Will climate change make some parts of the world much more difficult for survival? Will religious strife worsen so that safety is very difficult to achieve? Is Mary Poppins or William Gibson the better prognosticator of what the world will look like in thirty years? A plan that looks appealing in a Mary Poppins world may look much worse in Gibson's anti-utopian city of the future (Gibson 1986).

And then there is the difficult question of *akrasia*—weakness of the will. Can I successfully carry out my long term plans? Or will short term temptations make it impossible for me to sustain the discipline required to achieve my long term goals? For that matter, how much should future goods matter to me in the present?[2]

There are several features of a full human life in prospect that make it difficult to formulate a satisfactory theory of the formulation and assessment of rational life plans.

- The extended timeframe of the planning problem: formulating a plan in one's twenties that is intended to guide through the end of one's life in his or her nineties.
- The fact of a person's plasticity. Features of character, personality, habit, taste, and preference are all subject to a degree of purposive change. So it would seem that these should be the object of rational deliberative planning as well. But it is hard to see how to do this.
- The fact of the unpredictability of the external environment, both natural and social.
- The difficulty of designing a plan that is robust through dramatic change within the person.
- The difficulty of incorporating possible future capabilities of changing the self and the body directly through choice.

These challenges make traditional rational-choice theory unpromising as a foundation for arriving at a theory of life planning. Traditional rational choice theory is designed around the assumption of exogenous and fixed preferences, the ability to assign utility to outcomes, and quantifiable knowledge of the likelihood of various outcomes. But the five factors mentioned here invalidate all these assumptions. We may want to include features of character and preference in the things we would like to change through our lives, and we know there is great uncertainty about the features of the natural and social environment. It is simply impossible to look at life planning through the lens of the maximizing-expected-utility framework. So rational choice theory is useless in this context.

Deliberation

The idea of a "rational life plan" highlights "rationality," which has the connotation of short-term, one-off decision making. And this implication plainly does not fit the problem of life planning very well. Living a life is more like the making of a great sculpture than it is planning a Napoleonic military campaign. The sculptor makes unanticipated adjustments in response to the revealed characteristics of the stone. But what if we shifted the terms of the question and asked instead, what is involved

---

[2]Jon Elster looks at weakness of the will as a collective action problem across stages of the self in "Weakness of Will and the Free-Rider Problem" (Elster 1985). He considers the broader topic of imperfect rationality in *Ulysses and the Sirens*; Elster (1979).

in being *deliberative and reflective* about the direction of one's life? Does this give more room for bringing the idea of rationality into the idea of a life plan? It turns out that broadening our focus from calculation about costs and benefits to deliberation about values and goals improves our ability to formulate a theory of "rational life plans"; and it brings our account into much closer proximity to the pragmatist theory of action to be described below.

Being deliberative invokes the idea of considering one's goals reflectively and in comparison, considering strategies and actions that might serve to bring about the realization of these goals, and an ongoing consideration of the continuing validity of one's goals and strategies. Instrumental rationality takes a set of goals as being fixed; deliberative rationality works on the assumption that it is possible to reason reflectively about one's goals themselves. This is the thrust of Socrates' "unexamined life"—the good life requires reflection and deliberation about the things one seeks to achieve in life. Here is how Aristotle describes deliberation in the *Nicomachean Ethics*, book 3 (Aristotle 1987):

> We deliberate about things that are in our power and can be done; and these are in fact what is left. For nature, necessity, and chance are thought to be causes, and also reason and everything that depends on man. Now every class of men deliberates about the things that can be done by their own efforts. And in the case of exact and self-contained sciences there is no deliberation, e.g. about the letters of the alphabet (for we have no doubt how they should be written); but the things that are brought about by our own efforts, but not always in the same way, are the things about which we deliberate, e.g. questions of medical treatment or of money-making. And we do so more in the case of the art of navigation than in that of gymnastics, inasmuch as it has been less exactly worked out, and again about other things in the same ratio, and more also in the case of the arts than in that of the sciences; for we have more doubt about the former. Deliberation is concerned with things that happen in a certain way for the most part, but in which the event is obscure, and with things in which it is indeterminate. We call in others to aid us in deliberation on important questions, distrusting ourselves as not being equal to deciding. (Nicomachean Ethics, book 3)

What Aristotle focuses on here is choice under conditions of uncertainty and complexity. Deliberation is relevant when algorithms fail—when there is no mechanical way of calculating the absolutely best way of doing something. And this seems to fit the circumstance of planning for a life or career.

How does "deliberation" come into the question of life plans? It is essential.

(1) The goals a person pursues in life cannot be specified exogenously; rather, the individual needs to consider and reflect on his or her goals in an ongoing way. Aristotle was one of the first to reveal that often the goals and goods we pursue are, upon reflection, derivative from some more fundamental good. But Kant too had a position here, favoring autonomy over heteronomy (Kant 1956). Reflection allows us to gain clarity about those more fundamental goods that we value.

(2) The particular strategies and means that we choose at a moment in time may have only a superficial correspondence to our more fundamental goods and values that is undercut by more rigorous examination. We may find that a given mode of action, a strategy, may indeed lead to good X, but may also defeat the

achievement of Y, which we also value. So deliberative reflection about the strategies and actions we choose can allow us to more fully reconcile our short-term strategies with our long-term goals and goods.

Economists and philosophers have sometimes maintained that values and goals do not admit of rational consideration. But this is plainly untrue. At the very least it is possible to discover positive and negative interactions among our goals and desires—the desire to remain healthy and the desire to eat ice cream at every meal are plainly in conflict. Less trivially, the goal of living life in a way that is respectful of the dignity of others is inconsistent with the goal of rising to power within a patriarchal or racist organization. It is possible that there are values that are both fundamental and incommensurable—so that rational deliberation and reflection cannot choose between them. But it is hard to think of examples in which this kind of incommensurability arises as a practical problem.[3]

Pragmatist Theory of Action

Rational choice theory is one possible approach to thinking about deliberate action. This involves analyzing the question of rationality from the point of view of means-end calculation and deliberation. However, pragmatist thinkers have offered a fundamentally different way of conceptualizing the relationships between action and thought. These approaches involve a number of important ideas that are orthogonal to means-end rationality—habit, creativity, and improvisation—while maintaining commitment to the idea that human action is intentional, thoughtful, and adaptive. George Herbert Mead, John Dewey, and Charles Sanders Peirce contributed to these theories in the early part of the twentieth century, and a number of contemporary sociologists have turned to these ideas in order to arrive at a more adequate theory of the actor for sociology. So what might a pragmatist theory involve?

Hans Joas has contributed extensively to the question of how pragmatism intersects with sociology.[4] Joas's survey article with Jens Beckert (Joas and Beckert 2001) on action theory is a good place to start, since Joas and Beckert are specifically concerned in that article to give an exposition of a theory of action that acknowledges several important sources for such a theory while specifically developing a pragmatist account.

Joas and Beckert begin their account by framing the standard assumptions of existing action theory in terms of two poles: action as rational choice (e.g. James Coleman) and action as conformance to a set of prescriptions and norms (e.g.

---

[3]Hilary Putnam and Vivian Walsh provide vigorous arguments against the separation of facts and values in economics; Putnam and Walsh (2012).

[4]Key contributions include Joas's books, *The Creativity of Action* and *Pragmatism and Social Theory* (Joas 1993, 1996).

Durkheim, Parsons). They argue for a view that is separate from both of these, under the heading of "creative action".

Common to both traditional views, Joas and Beckert argue, is the assumption of a strong interpretation of purposiveness: that action proceeds to bring about explicit pre-articulated goals subject to antecedently recognized constraints. The pragmatist view of action rejects this separation between goals, action, and outcome, and focuses on the fact that goals and actions themselves are formulated within a dynamic and extended process of thought and movement. (Dewey is the chief source of this view.) Tactics, movements, and responses are creative adaptations to fluidly changing circumstances. The basketball player driving to the basket is looking to score a goal or find an open teammate. But it is the rapid flow of movement, response by other players, and position on the floor that shapes the extended action of "driving for a layup." Likewise, a talented public speaker approaches the podium with a few goals and ideas for the speech. But the actual flow of ideas, words, gestures, and flourishes is the result of the thinking speaker interacting dynamically with the audience. Joas and Beckert put their view in these terms:

> At the beginning of an action process goals are frequently unspecific and only vaguely understood. They become clearer once the actor has a better understanding of the possible means to achieve the ends; even new goals will arise on the basis of newly available means. (Joas and Beckert 2001: 273)

> For the theory of creativity of action the significance of the situation is far greater: Action is not only contingent on the structure of the situation but the situation is constitutive of action. (Joas and Beckert 2001: 274)

So what are the features of the situation that intersect with the thinking actor to create the temporally extended action? Joas and Beckert refer to corporality and sociality. The body is not simply the instrument of the agent. Rather, the physical features and limitations of the body themselves contribute to the unfolding of the action. And the other persons involved in an action are not simply subjects of manipulation. Their own creativity in movement and action defines the changing parameters of the actor's course of action.

Neil Gross extends some of the insights from pragmatism to help us to formulate a more adequate framework of sociological thinking (Gross 2010). Against the hyper-calculative conception associated with rational choice theory, Gross (and pragmatism) advocates for a more fluid, interactional, and only partially conscious flow of actions. There is a suggestion here of stylized modes of behavior (scripts) within which persons locate their actions, and a suggestion of the importance of specific cognitive fields embodied in social groups that contextualize and rationalize the person's activities (assumptions, for example, of how a doctor should treat a patient in a hospital).

What makes this set of assumptions a "pragmatist" approach? Fundamentally, because it understands the actor as situated within a field of assumptions, modes of behavior, ways of perceiving; and as being stimulated to action by "problem situations". So action is understood as the actor's creative use of scripts, habits, and cognitive frameworks to solve particular problems. (Gross refers to this as an A-P-H-R chain: actor, problem situation, habit, and response; Gross 2010: 343).

But of course, not all action is habitual. The opposite end of the spectrum includes both deliberation and improvisation. These categories themselves are different from each other. Deliberation involves explicit consideration of one's goals, the opportunities that are currently available within the environment of choice, and the pro's and con's of the various choices.

Improvisation differs from both habit and deliberation. Improvisation is a creative response to a current and changing situation. It involves intelligent, fluid adaptation to the current situation, and seems more intuitive than analytical. The skilled basketball player displays improvisational intelligence as he changes his dribble, stutter-steps around a defender, switches hands, and passes to a teammate streaking under the basket for the score. At each moment there are shifting opportunities that appear and disappear as defenders lose their man, teammates slip into view, and the shot clock winds down. This series of actions is unplanned but non-habitual, and it displays an important aspect of situational intelligence. Bourdieu captures much of this aspect of intelligent behavior in his concept of *habitus* (Bourdieu 1977).

This account of action refrains from assuming a fully planned and calculating approach to decision-making by the individual. But neither do we get the opposite extreme—robots playing out their scripts without intelligence or adaptation to circumstance. Instead, each of the players in this story retains his/her own assessment of what is currently going on and what deviations of script may be demanded. The individual retains the ability and the responsibility to break with procedures when there is an imminent reason to do so. So the routines and scripts guide rather than generate the behavior.

This is a rich and nuanced theory of action, and one that has the potential for offering a basis of a much richer analysis of concrete social circumstances than we currently have. At the same time it should not be thought to be in contradiction to either rational-deliberation or normative-deliberation theories. These creative actors whom Gross and Joas and Beckert describe are purposive in a more diffuse sense, and they are responsive to norms in action. The most important point here is the contrast that the new pragmatists offer between stylized, mono-stranded models of action, and thick theories that incorporate the plain fact of intelligent adaptation and shaping of behavior that occurs in virtually all human activities.

So the pragmatist interpretation of social action is not inherently inconsistent with the idea of intelligent, purposive action. Instead, we can think of the actor in this case as involving in a complicated series of behaviors that reflect both deliberation and internalized script. This interpretation is analogous to Pierre Bourdieu's position on the subject of habitus in *Outline of a Theory of Practice* (Bourdieu 1977). Conduct that is guided by norms (in this case, scripts and roles) can nonetheless also be intelligent and strategic. Seen in this light, the pragmatist interpretation supplements the purposive theory of action rather than replaces it.

This account of action appears to have a great deal of relevance to the topic at hand—"purposiveness in creating a life". Rather than imagining that a life plan is a blueprint and a well-lived life is one that unfolds according to this prior design, we may instead emphasize the qualities of improvisation and directedness that are at the heart of the pragmatist theory of action. This way of framing "living a life intentionally" seems to make a great deal of sense in the biographies of highly accomplishing individuals, from Paul Dirac (Farmelo 2009) to Dietrich Bonhoeffer (Metaxas 2010).

Bounded Life Planning

It is worth asking whether rational life plans actually exist for anyone. Perhaps most people's lives take shape in a more contingent and event-driven way. Is it possible that a person's journey takes its shape in a more stochastic way: opportunities, short term decisions, chance involvements, and some ongoing efforts to make sense of it all in the form of a developing narrative? Here we might say that life is not planned, but rather built like Neurath's raft with materials at hand; and that rationality and deliberation come in only at a more local scale.

This point makes quite clear the relevance of pragmatist theories of action for the way we think about a person's deliberativeness and construction of his or her life. Perhaps guided opportunism is the best we are likely to do: look at available opportunities at a given moment, pursue the opportunity that seems best or most pleasing at that point, and enjoy the journey. Or perhaps there are some higher-level directional rules of thumb—"choose current options that will contribute in the long run to a higher level of X". In this scenario there is no overriding plan, just a series of local choices. This alternative is pretty convincing as a way of thinking about the full duration of a person's life, as any biographer is likely to attest.

On this approach we might think of life planning as being less like a blueprint for action and more like a navigational guide. We might think of the problem of making intermediate life choices as being guided by a compass rather than a detailed plan—the idea that we do good work on living if we guide our actions by a set of directional signals rather than a detailed map. Life outcomes result from following a compass, not moving towards a specific GPS point on a map.

Here, then, is a simple way of characterizing purposive action over a long and complex period. The actor has certain guiding goals he or she is trying to advance. It is possible to reflect upon these goals in depth and to consider their compatibility with other important considerations. This might be called "goal deliberation". These goals and values serve as the guiding landmarks for the journey—"keep moving towards the tallest mountain on the horizon". The actor surveys the medium-term environment for actions that are available to him or her, and the changes in the environment that may be looming in that period. And he or she composes a plan for these circumstances—"attempt to keep moderate Southern leaders from supporting secession". This is the stage of formulation of mid-range strategies and tactics, designed to move the overall purposes forward. Finally, like Odysseus, the actor seizes unforeseen opportunities of the moment in ways that appear to advance the cause even lacking a blueprint for how to proceed.

We might describe this process as one that involves local action-rationality (pragmatist theory) guided by medium term strategies (plans) and oriented towards long term objectives (deliberation about values and objectives). Rationality comes into the story at several points: assessing cause and effect, weighing the importance of various long term goals, deliberating across conflicting goals and values, working out the consequences of one scenario or another, etc. But equally important are the qualities of spontaneity, improvisation, and adaptation, as the individual recognizes and incorporates the value that an unexpected opportunity may have in facilitating the achievement of longer-term goals and aspirations.

As biologists from Darwin to Dawkins have recognized, the process of species evolution through natural selection is inherently myopic. Long term intelligent action is not so, in that it is possible for intelligent actors to consider distant solutions that are potentially achievable through orchestrated series of actions— plans and strategies. But in order to achieve the benefits of intelligent longterm action, it is necessary to be intelligent at every stage—formulate good and appropriate distant goals, carefully assess the terrain of action to determine as well as possible what pathways exist to move toward those goals, and act in the moment in ways that are both intelligent solutions to immediate opportunities and obstacles, and have the discipline to forego short term gain in order to stay on the path to the long term goal. But, paradoxically, it may be possible to be locally rational at every step and yet globally irrational, in the sense that the series of rational choices lead to an outcome widely divergent from the overriding goals one has selected.

I have invoked a number of different ideas here, all contributing to the notion of rational action over an extended time: deliberation, purposiveness, reflection, spontaneity, adaptation, improvisation, calculation of consequences, and intelligent problem solving. What is interesting is that each of these activities is plainly relevant to the task of "rational action"; and yet none reduces to the other. In particular, rational choice theory cannot be construed as a general and complete answer to the question, "what is involved in acting rationally over the long term?".

This approach has more in common with Herbert Simon's 1957 concept of bounded rationality and satisficing rather than maximizing as a rule of rational decision-making (*Models of Man*; Simon 1957). Simon's most celebrated idea was the notion of "satisficing" rather than "optimizing" or "maximizing" in decision-making; he put forward a theory of ordinary decision-making that conformed more closely to the ways that actual people reason rather than the heroic abstractions of expected utility theory. This idea is highly relevant to the topic of life planning. Instead of heroically attempting to plan for all contingencies over the full of one's life, a bounded approach would be to consider short periods and make choices over the opportunity sets available during those periods. And if we superimpose on these choices a higher-level set of goals to be achieved—having time with family, living in conformity to one's moral or religious values, gaining a set of desired character traits—then we might argue that this decision-making process will be biased towards outcomes that favor one's deeper values as well as one's short-term needs and interests.

This approach will not optimize choices over the full lifetime, as Rawls proposed; but it may be the only approach that is feasible given the costs of information gathering and scenario assessment. So what about a rational life plan? At this point the phrase seems inapt to the situation of a person's relationship to his or her longterm life. A life is more of a concatenation of a series of experiences, projects, accidents, contingencies—not a planned artifact or painting or building. A life is not a novel, a television series, or a mural with an underlying storyboard in which each element has its place. And therefore it seems inapt to ask for a rational plan of life. Individuals make situated and bounded deliberative decisions about specific issues. But they don't plot out their lives in detail.

What seems more credible is to ask for a framework of navigation, a set of compass points, and a general set of values and purposes that get invested through projects and activities. The idea of the *bildungsroman* seems more illuminating— the idea of a young person taking shape through a series of challenging undertakings over time. Development, formation, values clarification, and the formation of character seem more true to what we might like to see in a good life than achieving a particular set of outcomes.

Where, then, do thinking and reasoning come into the picture? This is where Socrates and Montaigne seem to be relevant. They look at living as an opportunity for deepening self-knowledge and articulation of values and character. Recall their most famous aphorisms: "To philosophize is to learn how to die" (Montaigne) and "The unexamined life is not worth living" (Socrates). The upshot of these aphorisms seems to be this: reasoning and philosophizing allow us to probe, question, and extend our values and the things we strive for. And having examined and probed, we are also in a position to assess and judge the actions and goals that are presented to us at various stages of life. How does a college major, a first job, a marriage, or a parenting challenge frame the future into which the young person develops? And how can practical reflection about one's current values help to give direction to the future choices he or she makes later in life?

Practical rationality perhaps amounts to little more than this when it comes to constructing a life: to consider one's best understanding of the goods he or she cares most about, and acting in the present in ways that shape the journey towards a future that better embodies those goods for the person and his or her concerns.

These considerations make it apparent that there is something wildly unlikely about the idea of a developed, calculated life plan in the sense described by John Rawls. Here is a different way of thinking about this question, framed about directionality and values rather than goals and outcomes. We might think of life planning in these terms:

- The actor frames a high-level life conception—how he/she wants to live, what to achieve, what activities are most valued, what kind of person he/she wants to be. It is a work in progress.

- The actor confronts the normal developmental issues of life through limited moments in time: choice of education, choice of spouse, choice of career, strategies within the career space, involvement with family, level of involvement in civic and religious institutions, time and activities spent with friends, ... These are week-to-week and year-to-year choices, some more deliberate than others.
- The actor makes choices in the moment in a way that combines short-term and long-term considerations, reflecting the high-level conception but not dictated by it.
- The actor reviews, assesses, and updates the life conception. Some goals are reformulated; some are adjusted in terms of priority; others are abandoned.

This picture looks quite a bit different from more architectural schemes for creating and implementing a life plan considered by more rationalist philosophers, including the view that Rawls offers for conceiving of a rational plan of life. Instead of modeling life planning after a vacation trip assisted by a tourist agency guide-book (turn-by-turn instructions for how to reach your goal), this scheme looks more like the preparation and planning that might have guided a great voyage of exploration in the sixteenth century. There were no maps, the destination was unknown, the hazards along the way could only be imagined. But there were a few guiding principles of navigation—"Keep making your way west," "Sail around the biggest storms," "Strive to keep reserves for unanticipated disasters," "Maintain humane relations with the crew." And, with a modicum of good fortune, these maxims might be enough to lead to discovery.

This scheme is organized around directionality and regular course correction, rather than a blueprint for arriving at a specific destination. And it appears to be all around a more genuine understanding of what is involved in making reflective life choices. Fundamentally this conception involves having in the present a vision of the dimensions of an extended life that is specifically one's own—a philosophy, a scheme of values, a direction-setting self understanding, and the basics needed for making near-term decisions chosen for their compatibility with the guiding life philosophy. And it incorporates the idea of continual correction and emendation of the plan, as life experience brings new values and directions into prominence.

The advantage of this conception of rational life planning is that it is not heroic in its assumptions about the scope of planning and anticipation. It is a scheme that makes sense of the situation of the person in the limited circumstances of a par-ticular point in time. It doesn't require that the individual have a comprehensive grasp of the whole—the many contingencies that will arise, the balancing of goods that need to be adjusted in thought over the whole of the journey, the tradeoffs that are demanded across multiple activities and outcomes, and the specifics of the destination. And yet it permits the person to travel through life by making choices that conform in important ways to the high-level conception that guides him or her. And somehow, it brings to mind the philosophy of life offered by those great philosophers of life, Montaigne and Lucretius.

Ironically, it may be that a rational plan of life is more like a narrative within which a person's actions and choices make sense than it is to a set of blueprints for a future building, and further, that this narrative is more compelling in the later stages of life than in youth. On this interpretation, the individual creates a life through his or her thoughts, deliberations, and choices in the moment, and these choices eventually add up to a meaningful narrative. And that narrative can only be understood fully in the later part of the individual's history. This is a version of Hegel's theory of the "cunning of reason" in the preface to *The Philosophy of Right*: the rationality of a life can only be discerned at the end, "only when philosophy paints its grey-on-grey" (Hegel 1967[1821]).

These considerations make two things fairly clear. First, the idea of a life plan is a metaphor rather than a realistic description of the way that deliberative and reflective people organize their lives. Indeed, it is a bad metaphor, because it leads us to overlook the inherent contingencies, both internal and external, which a life involves. And second, these reflections seem to show that the "life plan fundamentalism" associated with Aristotle and Rawls is misguided.

References

Aristotle, (1987). *The nicomachean ethics*. Buffalo, N.Y.: Prometheus Books.

Baron, J. (1985). *Rationality and intelligence*. Cambridge: Cambridge University Press.

Bourdieu, P. (1977). *Outline of a theory of practice*. Cambridge: Cambridge University Press.

Bratman, M. (1987). *Intention, plans, and practical reason*. Cambridge, MA: Harvard University Press.

Elster, J. (1979). *Ulysses and the sirens: Studies in rationality and irrationality*. Cambridge [Eng.]; New York: Cambridge University Press.

Elster, J. (1985). Weakness of will and the free-rider problem. *Economics and Society, 1*, 231–265. https://doi.org/10.1017/S0266267100002480.

Farmelo, G. (2009). *The strangest man: The hidden life of Paul Dirac, mystic of the atom*. New York: Basic Books.

Gibson, W. (1986). *Neuromancer* (1st Phantasia Press ed.). West Bloomfield, MI: Phantasia Press.

Gross, N. (2010). Charles tilly and American pragmatism. *The American Sociologist, 41*(4), 337–357.

Hegel, G. W. F. (1967 [1821]). *The Philosophy of right*. In T. M. Knox (Ed.). London, New York: Oxford University Press.

Joas, H. (1993). *Pragmatism and social theory*. Chicago: University of Chicago Press.

Joas, H. (1996). *The creativity of action*. Cambridge: Polity Press.

Joas, H., & Beckert, J. (2001). Action theory. In J. H. Turner (Ed.), *Handbook of sociological theory* (pp. 269–285). New York: Kluwer Academic.

Kahneman, D., Slovic, P., & Tversky, A. (1982). *Judgment under uncertainty: Heuristics and biases*. Cambridge: Cambridge University Press.

Kant, I. (1956). *Critique of practical reason*. New York: Liberal Arts Press.

Metaxas, E. (2010). *Bonhoeffer: Pastor, martyr, prophet, spy: A righteous gentile vs. the third Reich*. Nashville: Thomas Nelson.

Putnam, H., & Walsh, V. C. (2012). *The end of value-free economics, Routledge INEM advances in economic methodology*. London, New York: Routledge.

Rawls, J. (1971). *A theory of justice*. Cambridge, MA: Belknap Press of Harvard University Press.

Simon, H. (1979). From substantive to procedural rationality. In F. Hahn & H. Martin (eds.) *Philosophy and economic theory*.

Simon, H. (1957). *Models of man: Social and rational*. New York: Wiley.

von Neumann, J., & Morgenstern, O. (1944). *Theory of games and economic behavior*. Princeton: Princeton University Press.

Chapter 11
Dynamics of Rationality and Dynamics of Emotions

Pierre Livet

Abstract From a dynamical perspective, rationality implies to revise our beliefs if necessary. Understanding emotions needs to analyze the interaction between our internal dynamics and the dynamics of our environment. The combination of these two dynamical perspectives leads to take into account the impact of emotions in our decisions and the relations between emotions and revision. The different effects of two kinds of anticipating emotions (representational and driving ones) can explain biases of temporal discounting. Comparative and mixed emotions can explain biases in our choices under uncertainty. But recurrent emotions (in the long term) can also be incentives to revise reasonably our order of preferences—the preferences that resist this revision are our more stable values. As different processes of revision are possible, leading to different ways of merging different kinds of order, this raises the problem of how to choose between different orders. Emotional dynamics can lead to shift from one kind of order to another and to learn what is the more reasonable order. But emotional learning is path dependent and takes time to experience the different orders and reach stability.

Rationality presents different aspects: epistemic rationality, decision rationality, instrumental rationality and axiological rationality. Epistemic rationality governs the relations between propositions, inferences and beliefs. Decision rationality governs the relations between beliefs, desires or preferences and choices—in relation with consequences of actions. Instrumental rationality governs the relations between goals (determined in relation with decision rationality), actions and means. Axiological rationality does not really determine values, ends and aims, but examines whether different ends are compatible and what hierarchy between ends is more reasonable. Rationality in general is supposed to ensure reasonable relations between values or ends and evaluations of propositions, inferences, beliefs, preferences, choices and means for action.

P. Livet (✉)
CEPERC, Aix Marseille University, Aix-En-Provence, France
e-mail: pierre.livet@univ-amu.fr

As one can see, rationality is relational. A proposition is not rational in itself, but only in relation to other propositions, and similarly for inferences, beliefs, desires, goals and means. The relational property of coherence appears to be a common requirement to all these diverse relations: propositions have to be coherent together, inferences have to ensure coherence, beliefs have to be coherent with other beliefs, choices have to be coherent with beliefs and with evaluation of consequences in accordance with preferences, means have to be coherent or at least compatible between them and with goals, evaluations have also to be coherent with one another. Coherence is not guaranteed between all these domains (for example between evaluations and means), but it is the main norm of rationality—at least in a static account of rationality.

Dynamical Rationality

This static perspective is not sufficient. We have also to take into account the dynamical aspect of rationality. Static propositions would not be very useful without the dynamics of inferences that relate propositions with one another. Beliefs are not necessarily true, and false beliefs or past true beliefs that are revealed to be false at the present time have to be revised. In the same way, decisions that have been made conditionally to false beliefs have to be revised. Revision is a dynamical operation. Economists assume that preferences have not to be revised (they have to be restricted instead to particular times and situations), but accumulating bad consequences that cannot be imputed to false beliefs and bad means is a rational incentive to revise preferences. Means may have to be revised if they are not the good ones. In the same way, evaluations that lead systematically to bad choices and bad results have to be revised.

In this dynamical perspective, static coherence is no longer a sufficient property. The dynamics of the relations between our actions and environment opens the possibility that new information is coming, and a part of this information may appear to be incoherent with what was previously expected on a rational basis (rational beliefs and probabilities). Absolute coherence is then no longer possible. We have to acknowledge the possibility of incoherence between our expectations and new information. We need then to change at least some of the beliefs in order to restore the coherence of information with our beliefs (Schlechta 1997). We also need to change the beliefs that were responsible (via inferential combination with other beliefs) for our defeated expectations. It seems rational to minimize such change, but there are different ways of defining the minimization process. It could even be useful not to suppress some of the concerned beliefs, but only to encapsulate them in order to prevent their present influence on our inferences and choices in a particular situation, if we want to keep them in reserve for other kinds of situations. It is not necessarily rational to delete systematically any belief and mode of inference that cannot be demonstrated to preserve coherence in every context (contrary to what logical inferences are required to do).

One has to accept local and temporary violation of coherence if one wants to restore coherence on a longer term. But the impact of revision is not limited to the short term. The first revision that we had to make has shown us that we need sometimes to revise our beliefs because our expectations are defeated. The problem is that another sequence of new information could again lead us to revising our previous revision. If our inferences are non-monotonic[1] (revisable) the results of our revisions may still lead to defeated expectations and then to revising of our revisions. This could lead to the "pessimistic argument": since we find errors in past science, we have to expect errors in future science, and so on and so forth, without any hope of reaching a state in which science will ensure certainty. The argument is flawed, because a dynamics of revision cannot be said rational except if it bring about relative improvement at each step and therefore also at the particular step of a revision of the previous method of revision. Such evolution improves in the long term the validity of the majority of the revised expectations. Of course, one method of revision could be revealed to be bad, but other methods will correct some of the defects of the previous one, even if they will still have defects. Absolute certainty is not possible—this impossibility is a consequence of the dynamical perspective—but if we have revising methods, improvement in the long term is more likely than if we stick to our errors.

A more awkward problem is that different methods of revision are possible. The most popular scientific method of revision is Bayesian revision. It gives rules of revising priori probabilities given the observed probabilities that result from experiments. The main interest of the Bayesian rules is that their recursive application is proved to converge on a stable solution—after a very high number of experiments. Of course this long sequence of experiments takes a very long time and we have to assume that during all this time, the distribution of probability of the type of phenomenon that we are studying remains the same. For example, we assume that the "century wave" will occur ten times during the on-going millennium. But climate changes, and maybe these century waves will occur fifty times or more in our millennium. If we have reasons to believe that the change is rapid and important, maybe we need a less progressive and more rapid rule of revision. Non-monotonic revisions can lead to more rapid revisions, but they do not guarantee convergence on a solution that will be stable in the long term. As there is no unique solution to the revision problem, it is rational to examine the different methods and to try to define their respective conditions and domains of application. For example, we could use a rather drastic revision if our expectations have been suddenly massively defeated, and a Bayesian revision if we can assume that the evolution of the domain is not disruptive. Of course, our evaluation of the relevant

---

[1]A function is monotonic, if when it is increasing (respect. decreasing) it never decreases (respect. increase) when its independent variable increases (respect. decrease).

conditions is not guaranteed to be the good one. But when we will have to revise our evaluation, we will still use one of the revision methods that are in our repertory, and try to define new methods only if the method that best fits the circumstances is still not satisfying. No other strategy could be rational. The consequence is that dynamical rationality requires us to explore different revisions methods and their respective conditions of application, in particular their dynamical conditions and their mode of temporality—mainly whether the change is a progressive or an abrupt and disruptive one.

Emotional Dynamics

Emotions have also to be considered in a dynamical and not in a static perspective. Some theories of emotions (Solomon's theory, for instance) are static, as they identify emotions with evaluations and evaluative judgments. In these theories, fear is supposed to be reducible to the evaluation of something as dangerous in the present situation. But we can make evaluations without emotions; therefore emotions are not reducible to evaluations and evaluative judgments, even if they imply appraisal. Frijda (1986) claims that emotions are "action readiness potentials", that trigger feelings of our body's readiness to react to a property of the environment that may have a positive or negative impact on our situation (Deonna and Teroni 2012). But we can also feel emotions when we contemplate a landscape or a painting, with no perspective of action. The core of Frijda's idea may be that emotions are related to our dynamic—"readiness" can be understood as a tendency not to stay in the same situation but to trigger new dynamics. The dynamics of action consists in mobilizing energy for changing the current situation. The dynamics of contemplation consists in inhibiting action tendencies and exploring the perceptive information that we receive. In general, the subjective dynamics implies a modality of projection from the past and present towards the future, and an anticipation of the future (either as changing or as going on). We do not need to be explicitly aware of this anticipation, but anyway it guides our action or exploration and leads us to focus on some aspects of the situation.

Nevertheless the burst of occurrent emotions would not be possible to explain if we would only consider the anticipative dynamics, that I call our internal dynamics. Occurrent emotions need a difference between the internal dynamics and the dynamics of the environment. One may wonder whether this external dynamics has to be taken into account. One would say that as we are only sensitive to dynamics that are relevant for our own purposes, our internal dynamics seems sufficient to give an emotional color to whatever external change of the situation. But anticipative representations of how the external dynamics will look like are already parts of our internal dynamics.

In order to avoid misunderstandings, please note that I do not reduce the content of these anticipative representations to dynamics as a kind of flux and its force: this content involves reactivation of sensorial memories, imaginative representations, and evaluations. But emotions are raised by differences between this content as a dynamical anticipation and the content that we perceive that the dynamics of the environment provides. Therefore when the real external dynamics is the anticipated one, there is no emotional burst. It requires a difference between the real external dynamics (at least the features of the real external dynamics that we perceive) and the anticipative representation that is implicit—or explicit—in our internal dynamics. External dynamics can even activate internal potential dynamics that were not active before the impact of the change of the situation.

Of course the content that we assign to the external dynamics is appraised relatively to the anticipating perspective of our internal dynamics and to the newly activated internal dynamics; but this anticipating perspective cannot autonomously change itself in the way in which external dynamics impacts it. For example, when we imagine such impact, in order to feel emotions, we need to activate past similar features of external impacts in order to be moved by these emotions, and these activations have to differ from our current internal dynamics. Emotion is like perception in this respect: perception is richer than imagination and internal representations; occurrent emotion is richer than simple internal representations and imaginations.

Emotions have been selected during the evolution. Determining their evolutionary potential could help us to find an evolutionary rationalization of the role of emotions, if not the rationality of emotions. What is this potential? Emotions make possible diverse reactions that are adaptive to very different regimes of impact and different temporalities of the changes of external dynamics—from a very rapid and disruptive change with a huge impact on our plans of actions and welfare, to a minor inflexion relatively to our anticipating dynamics. For example fear may induce simple vigilance, or aggressiveness, or flight, or frozen behavior, in accordance with the diverse impacts of dangers. Emotional reactions are not guaranteed to be optimal, but they make us able to react to a large range of changes in external dynamics. Thanks to emotional memory and imagination, we can even react to small changes that occur in daily life when we know that such changes will have very important consequences in the long term.

Can we move from this evolutionary rationalization to the rationality of emotions? The problem is that the very variety of emotional reactions—which is linked with their adaptive power—is also a source of incoherence, when one modality that fits a particular regime of dynamics or temporality interferes with a modality that fits another regime. Let us compare this emotional problem with the problem of rational revision. At first sight, there is a huge difference between the rationalization of emotions that is grounded on their evolutionary reliability and the rationality of revision methods. The later make salient the problem of combining different methods, and researchers try to elaborate methods for dealing with this problem. On the contrary, emotions do not seem to lead to any way of solving the inconsistencies raised by combining different regimes of emotional reactions. I will first analyze some examples of these inconsistencies (temporal discount, and biases of choices under uncertainty) and then show that nevertheless an emotional treatment of this problem exists, that we need to pay attention to it and to reinforce its possible links with rational revisions. It will appear that there are emotional revision processes that they are incentives to revisions of beliefs and are even necessary incentives for revising preferences.

Temporal Discount and Anticipating Emotions

The temporal discount bias consists in preferring to get a smaller gain A in the nearer future to getting a bigger gain B in a more remote but still near future. It would be rational to prefer to get a given gain in a nearer future to the same gain in a very remote future, if we are uncertain about the remote future. But when the two gains are both in a very remote future we reverse this preference and prefer B to A. In this situation we choose to get the bigger and more remote gain. Reversing our preferences induces temporal inconsistency and irrationality.

Literature of theory of decision has not really try to explain this bias, but mainly to formalize it and define what is the curve of the temporal discount—hyperbolic curve or a combination of hyperbolic and exponential curves (Frederick et al. 2002; Han and Takahashi 2012). The relations with emotions have been taken into account only in few papers [see Geoffard and Lucchini (2010) for example]. I believe that a purely emotional account is possible, if we take into account the interaction between internal and external dynamics, and introduce the notion of "anticipating emotions".

When we evoke a possible future situation that differs from our present course of action and its internal dynamics, this evocation can trigger an emotional reaction that is similar to the memorized effect of the past interaction of similar external dynamics and similar past internal dynamics. These emotions are "anticipating emotions". Anticipating emotion is the result of an interaction between the internalization of a supposed future external dynamics and its feedback on our internal dynamics, which defines a new internal dynamics. For example, if we anticipate that we will not get a gain at $t + 2$ that we would have got if we would have made at $t + 1$ a different decision, this anticipation triggers anticipating regret.

Within the general category of anticipating emotions, we need to distinguish representational anticipating emotions and driving anticipating emotions. While representational anticipating emotion is mainly the emotional effect of a representation of the future situation that has no direct effect on our action, driving anticipating emotion is direct incentive to act or at least to put this action in our agenda (even if this incentive may be inhibited by other motivations). For example, anticipating that we could miss an opportunity may drive us to take it. Driving emotions, even if they are anticipatory emotions, are not active all the time, but mainly near the right time for acting.

The common sense explanation of temporal discount is impatience. We prefer getting a smaller gain A at time t_1 to getting a bigger gain B at time t_2 $(t_2 > t_1)$ because we are impatient to get a gain. But if impatience would only be related to differences between waiting times, the agent would also prefer A–B when located in a remote future (where A: t_3, B: t_4; $t_3 \gg t_2$ but $(t_4 - t_3) = (t_2 - t_1)$. Nevertheless, the difference between purely representational anticipating emotions and driving ones is one key for the solution of the problem, and impatience is merely the affect that related the two kinds of anticipating emotions.

Let us consider a paradigmatic temporal discount situation. At time t_0, I examine what to choose: either to get gain A at time $t + 1$, or gain B at time $t + 2$. A < B. Suppose that at time 0, I decide to wait for time $t + 2$ in order to get B. But suppose then that at time $t + 1$, I choose to get A, instead of waiting for getting B at time $t + 2$ (as most people do). What could be the emotional motivations for such choice?

We have to take into account the emotions raised by the interaction between our internal dynamics and the external dynamics during the period between t_0 and $t + 1$ and between period $t + 1$ and $t + 2$. Our basic internal dynamics consists in expecting to get either gain A or gain B. The general effect of positive or negative external dynamics on our experience of time is the following. If our expectations and internal dynamics are positive and the external dynamics is favorable during the waiting period, we experience time as shorter; if it is unfavorable, we experience time as longer. In the present case, we assume that external dynamics is neutral in the larger part of the $(t_0 - t + 1)$ period. But let us consider the smaller period from $(t + 1$ minus epsilon) to $(t + 1)$. It is related with the imminent event of not getting A. The anticipation of this event is a consequence of our initial choice in favor of B. This event presents an external unfavorable dynamics. Our anticipation of this external dynamics triggers another expectation, the expectation of our imminent regret of not getting A. This gives rise to an "anticipating emotion" of regret, which triggers the fear of missing the opportunity of getting A.

In this case, time between $t + 1$ and $t + 2$, considered from the perspective of $(t + 1$ minus epsilon), appears to be longer and more painful, because of the anticipated external dynamics polluted by anticipated regret; we compare it to the short and pleasant time $(t + $ epsilon$)$ that we would experience if we would choose to get A at $t + 1$. Of course we have to take also into account the positive anticipating emotion of getting the bigger gain B, which leads to experiencing time as shorter and pleasant. The two anticipating emotions are competing with one another.

The two representational anticipating emotions might counterbalance each other. But we have also to take into account the driving anticipatory emotion related to the closeness of the moment in which the action of getting gain A will be possible. When the decision is urgent, driving anticipatory emotions make our experienced time shorter than when time is experienced in relation with representation anticipatory emotions.

There are two decisive factors. The first one is that, at time $(t + 1$ minus epsilon), the pleasure of waiting for the (slightly) bigger gain B is appraised through the filter of the negative representational anticipating emotion of future regret and the attractiveness of getting A immediately. The second one is that getting A immediately is a component of the present external dynamics that gives rise to a driving anticipatory emotion. Seeing things through these two filters raises the threshold that our motivation for getting B has to exceed in order to drive us to wait for B instead of getting A at once. The driving anticipatory emotion related to getting A is a stronger incentive to action than the representational anticipatory emotion of getting B. Our anticipating emotion of getting the slightly bigger gain B will not be sufficient to overcome both the impact of our anticipating emotion of regret and the driving force of the attractiveness of the immediate opportunity of getting A. This emotional conflict may be a sufficient reason for preferring to get A.

When A and B are in the remote future $t + n$ (here n is big), we have to wait a long time, and our anticipating emotions would have to accumulate the different possible external dynamics along this long time. But all these anticipating emotions are only representational ones and not driving ones. In addition, we do not know what the relations between our internal and the external dynamics will really be at this remote time. Waiting from $t + n$ to $t + n + 1$ is negligible in comparison with the long wait between t_0 and $t + n$. There is no "$t + 1$ minus epsilon effect", as such time is also very remote. There is no effect of being at the verge of taking the opportunity to get A. Waiting for a reward during such long time needs compensation and this compensation is better given by the additional value of B.

Another observations (only apparently in tension with the previous ones) show that people that have to choose in which order they want to visit a boring person and a good friend prefer to visit first the boring person and second the good friend. Here again, we have to take into account the relations between internal and external dynamics at the different steps of the sequence of actions, but in this case, our anticipating emotions are only mainly driving ones, as the two visit are mandatory and are both parts of our agenda. We prefer to see the first visit and the associated temporary inhibition of the other visit as related to the advancement of the goals of our action program, as a step towards a better situation, than as a pure constraint— this decreases the unpleasant overestimation of time related to inhibition.

Comparative Anticipating Emotions

In "Rationality, neuroeconomics and mixed emotions" (Livet 2010) I have shown that the diversity of emotions can explain well-known biases of our choices under uncertainty, like Allais' Paradox and overestimations of small probabilities. It is well known that small probabilities and highest ones, including certainty, have a bigger weight than medium-size probabilities (Kahneman and Tversky 1984, 2000). The theory of ranked utility gives a formal account of the skewed curve of the weights of probabilities (this formal account uses "Choquet's capacities"-monotonic functions on the interval 0–1,) but it does not give a psychological explanation of the skewed curve.

What I called "mixed emotions" in this paper are anticipating emotions that mix, for example, the pleasure of getting in the future (when the results of the lottery will have been given) a bigger but risky gain in comparison with getting a lower but less risky gain. These emotions are comparative anticipating emotions. The pleasant emotion that I have just mentioned could be called "elating relief": we feel anticipated relief because it is a riskier choice, and this relief will makes us still happier to get this higher gain, so we feel anticipated elation. The dual emotion is "prudential pride". We anticipate that the result of lottery will be the lower gain, and that we will be proud to have resisted the attraction of the higher but risky gain and to have been cautious. For a given lottery these two comparative anticipating emotions are competing each other.

In Allais' Paradox, the lottery with a sure gain has the effect that prudential pride dominates elating relief, while in the lottery with two uncertain gains (the same gains than in the previous one, but with different probabilities and no certain gain), elating relief dominates prudential pride. This could be considered not as irrational biases, but as a rational comparison, if we take into account the effects of comparisons between mixed anticipating emotions.

But in order to explain other experimental results, we need to take into account two residual biases. The first one is that even cautious people are attracted by higher probabilities of gain—this effect of higher probabilities holds only in the positive domain of gains. The second one is a difference between gains and losses. If (elating) relief in gains is raised by a comparison between the less probable and bigger gain—a riskier and highly unexpected gain- and the more probable, highly expected but smaller gain, this is not true for losses. Let us admit that avoided losses are the equivalent of gains in the negative domain. Relief in losses cannot be raised by a comparison between an unexpected avoided bigger loss and a highly expected avoided smaller loss. The later cannot be disdained, as higher probabilities of avoiding losses are the only things that allow us to experience a feeling of control in the domain of losses. This feeling implies to have some control on the external dynamics. In gains, we may accept to have less control if the external dynamics may be more favourable. When we consider avoided losses we are not prone, in order to avoid a bigger loss, to give up the control on the external dynamics that is ensured by the higher probability of avoiding of a smaller loss.

Nevertheless we can feel relief in losses (not elation, but "greater relief"). Elation relief in gains is raised by the comparison between an exciting riskier (*unexpected*) high gain and an unexciting but less risky (*expected*) lower gain. Greater relief in losses is raised by a comparison between two combinations of relief and anxiety: relief and anxiety related to the bigger riskier (*unexpected*) avoided loss, and relief and anxiety related to a smaller and *also unexpected* and risky avoided loss.

We could say that the attractiveness of an uncontrolled favourable external dynamics—in contrast with more controlled but less favourable external dynamics —in the domain of gains is replaced in the domain of losses by the comparison between two external dynamics that are both out of control, one being more favourable than the other, as it avoids a bigger loss. Elating relief and greater relief are not symmetrical.

These asymmetries between gains and losses are not convenient for formalization, but they are psychologically meaningful and cannot be said to be irrational in themselves. The problem is that they may lead to inconsistencies between decisions in different situations—for example, when the same set of possible options is presented in a positive version, in terms of gains, and when it is presented in a negative version, in terms of losses. The same problem lurks in the cases of temporal discounting. I have given a plausible psychological explanation of temporal discounting, which may appear meaningful and not irrational. But still its emotional processes lead to inconsistencies, and people that are moved by such emotional processes play the role of suckers in Dutch book arguments.

Recurrent Emotions and Emotional Revision

We cannot be confident that purely non-affective rational cognitive processes could be sufficient for avoiding these inconsistencies, because the emotional processes that are the indirect causes of inconsistencies are still active under our rational computations and cannot be deactivated. We cannot compete with emotional processes other than by having recourse to other emotional processes. But are there other emotional processes which may help us to reduce these inconsistencies? One possible mechanism is the effect of "recurrent" emotions (Livet 2002, 2016). By "recurrent" I means similar emotions that reoccur in the long term in similar circumstances.

Let us take the example of Paul. His task in his office job is restrictively defined as a part (P) of a collective work. He would prefer a more innovative cooperation with motivated partners (C) and he ranks even more highly to have more freedom and be responsible on his own (I) for a task taken as a whole. His order of priorities is therefore I > C > P. Then, from time to time, in the long term, he experiences responsibility on his own. On every such occasion, he recurrently feels stress because he bears alone the burden of responsibility. This is what I call emotional "recurrence". As a result, he becomes less eager to work on his own and ranks the cooperative tasks more highly. His order is now C > I > P. His negative recurrent emotions have acted as incentives to change his priorities.

Recurrent emotions, recurrent stress in the example, are emotions that occur repeatedly, but at sufficiently long intervals. This long-term condition ensures that emotion reoccurs after a relaxation time and a reinstallation of the expectations linked with the previous internal dynamics. As the effect of the external dynamics is recurrent, it impacts not only the present internal dynamics and its expectations but also challenges the adaptive power of reinstalling such expectations (one traumatic or ecstatic event can have the same effect as a recurrent emotion). Recurrence of (mainly negative) emotions in the long term can lead us to revise our expectations (Livet 2002, 2016).

Emotional recurrence is then an incentive to revise our expectations and change the order of priority between different expectations. Recurrent emotions act as internal dynamics that can change the priority of other internal dynamics. They act as internalized external dynamics. When they lead to revision of expectations, they change the relation between internal anticipative dynamics and external dynamics.

Suppose that in one domain of activities one have repeatedly chosen action A instead of B-for example buying one DIY (do it yourself) tool because it was almost immediately available instead of waiting a bit for a better tool. If one has been disappointed recurrently by the defects of such DIY tools, the internalization of the external dynamics leads to regret this impulsive behaviour and will change the relations between internal and external dynamics. In a new similar situation, anticipation of choosing A will be associated with anticipated regret of missing B because one would have been too easily attracted by A (here regret is an emotion related to guilty), in conflict with the anticipation of the disadvantage of missing A opportunity. As a consequence, resistance to the impulsion triggered by the second anticipation becomes justified by the expectation of avoiding regret, an anticipating emotion that has now priority.

Emotional recurrence is needed here, because it is difficult to change our expectations at will. One has to wait that the relations between internal and external dynamics have changed; these relations have affective components that require emotional incentive in order to be revised. Nevertheless the range of impact of these revisions is limited. Emotional revisions have not a general impact—except in the case of very strong traumatisms. Most of the time, changes are limited to the domain of activities in which emotions have reoccurred.

Recurrent emotions are only "incentives" to revise the order of priorities of our expectations. Even if they are by the way incentives to revise our evaluations, emotional revisions are not all-powerful. They do not have the power of revising the priority of our most entrenched expectations. As a consequence, they do not necessarily produce adaptive preferences. They are no more needed in order to revise superficial and temporary expectations—which can be revised in the shorter term by habituation. They play the role of filters of values, tests of the relative importance of the values that are related to our different expectations. As evaluative judgements are also subject to illusion about what are our real values, emotional revisions reveal what values are our real priorities, even if the impact of such revisions is usually restricted to the context of the recurrent emotions.[2]

Nevertheless, recurrent emotions may also be efficient for changes that have the power to impact different domains of activities. Let us assume that each specific recurrent emotion is related to a particular ordering in a domain of expectations. Usually, we call "values" the properties or qualities that make possible this ordering. Examples of domain of expectations could be friendship relations, love relations, social status relations, work, leisure, and the like. Values would be friendship, love and tenderness, social recognition and prestige, efficiency and proficiency, fun, etc. Revising the expectations of one domain may have a negative impact, but also, in different conditions, a positive one on our order of preferences in another domain expectations.

For example, consider a manager who prefers to become a cook. When she works as a manager, her professional expectations about high income are satisfied, but her desire and expectations in other domains, like having friends, being independent, exercising manual talents and showing immediately useful creativity are

---

[2]As a reviewer suggests, such revision could be related to the aristotelician notion of « mediety », because extremist positions are likely to be discarded by revision that mixes different orders.

defeated. She feels recurrent stress and negative emotions like anxiety, envy, rivalry and the like. In addition, let us assume that her expectations about promotions (and still higher salary) have been recurrently defeated. The recurrence of negative emotions related to growing income and prestige in the professional domain can lead her to revise the priority of this domain of expectations, to give priority to aspects of other domains (independence, useful manual creativity, friendly relations with customers) and to move from the job of manager to the job of cook. In the domain of expectations of income and prestige, this is decline; in the domain of autonomous creativity and in the domain of friendly relations, this is a progress. Her move is rational if the overall evolution across her different domains of expectations is positive.

Now suppose that she evaluates her evolution from the point of view of her revised priorities, the ones of cook. What ensures her that such revision is justified, that it is the right revision across different domains of expectations?

Recurrent *positive* emotions have a role here. Suppose that our manager moves to a job of cook because she has a breakdown as manager but that she feels no saliently positive satisfaction in cooking. In this situation, revising the priority of the domain of expectations about income and professional prestige appears as a case of purely adaptive preferences. On the contrary, if she feels recurrently (not only in the short but in the long term) strong satisfaction as a cook, her move to this new job will be justified, because there is no emotional incentive to revise her previous revision.

Emotional Revision and the Problem of Fair Allocation

It appears that emotional revision can be seen as related to the emergence of an *intra-personal* order from different orders in different domains of expectations inside the agent's mind. The problem of finding a sensible and emotionally stable ordering between different orders in different domains of expectations exhibits rather strong analogies with the problem of extracting an *inter-personal* order of a collective choice of allocations, namely the problem of "fair allocations". In this problem, different transfers, different revisions are possible. Fair allocation implies a second order revision of some of these possible revisions, which restores collective equity and ensures a collective fair social order.

More formally, this similarity between revision and collective rule of redistribution is not a surprising thing. We already know that there is a quasi-translation from the different axioms of social choice to the different rules of non-monotonic logics (Rott 2001, 2002; Schelchta 2004) that are used in order to formalize the revision of beliefs.

In order to find a fair social order as well as to integrate different incentives to revise preferences, one has to extract a higher-level order from the different orders. In the literacy of theories of revision, such a problem is known as the problem of merging together different orders.

Merging order 1 and order 2 implies to intercalate rank ri in order 2 between ranks rj and rk in order 1. If rk has a high priority of order 2, it will have a lower priority in the merged order. Seen from the point of view of rk, merging does not satisfies monotonicity, even if the order between rank ri and rj of order 2 is still respected, as well as the relative order of the ranks of order 1. But the merged order, of course, respects monotonicity.

In the same way, the transfers needed for defining a fair allocation may require decreasing the higher level of income of richer people, in order to raise the lower level of poorer people. If we follow the maximin rule (respect. the more sophisticated leximin), the collective allocation gives priority to poor (respect. to poorer people at each level of comparison). For some richer people, such a change is non monotonic, while for poor it is monotonic, and by the definition of the rule the collective change is monotonic (since fairness implies here to strictly prefer collective allocation A to collective allocation B only if the final income of poor in 1 is better than in B.)

Can we elaborate in a more systematic way an analogy between the transfers required by a fair allocation and the (second order) revision needed by the quest for a remedy to emotional temporal inconsistencies and able to resist the succession of emotional revisions (triggered by different recurrences of different negative emotions)?

We have seen that recurrent emotions may in some cases lead us to change the priority of a domain of expectations relatively to the rank of another domain. But are there rational procedures that govern this change?

Fleurbaey and Maniquet's theory of fair allocation (2012) proposed to use indifference curves.[3] They give the example of indifference curves between two goods, wine and bread. Wine is here supposed to be substitutable to bread, and conversely, of course inside some limits. The evaluation of the *preference* for one amount w of wine rather than another amount w' may depend not only on w an w', but also on the amounts of bread b and b' that could be substituted to amount w or w' of wine, and on the total available amount of bread and wine.

This last amount is used to compare different indifference curves, each presenting a different set of combinations of amounts of wine and bread. Each indifference curve is representative of a given agent. The scale for such a comparison is given by the total endowment of wine and bread that is available, which defines the relative scarcity of wine relatively to bread.

What would be the analogues of such components and structure in the emotional sphere? Of course I do not claim that we can use all the formal properties of indifference curves in the domain of emotions (cardinality would be disputable, as well as differentiability), but only that the relational properties that indifference curves are assumed to represent may have conceptual analogues in this domain.

---

[3]They use indifference curves as a trick for relaxing the condition of Independence of Irrelevant Alternatives (restriction of evaluation to comparing items by pairs, without taking into account their relations to other items) that leads to Arrow's Impossibility Theorem.

Formally, we may shift from cardinality to orders or pre-orders. In the problem of inter-personal fair allocation of goods between different agents, each indifference curve represents the substitution ratios of different amounts of one good to different amounts of another one for a given agent. In our intra-personal problem of fair emotional revision of priorities between different domains of expectations for one given agent, the analogue of one indifference curve would be a given domain of expectations. In this domain of expectations, for example professional success in one's work, one does not mind whether during his work one gets satisfaction from one or from another aspect of what one assumes to be valuable in this domain, for example prestige or nice collaborations. If one gains prestige, one is not emotionally affected by moderately difficult collaborations, and conversely. More prestige compensates for less collaboration. In these situations, negative recurrent emotions do not trigger revisions, as positive emotions compensate for negative emotions. These compensations are the analogue of the substitution ratios of an indifference curve. As positive emotions compensate for negative emotions, there are no emotional incentives for revising the priority of this domain of expectations relatively to other domains.

Different indifference curves may cross each other in one quadrant (let say, the positive one) for example when one curve starts from a lower point but with a steeper slope and cross another one that starts from a higher point but with a gentle slope. The emotional analogue would be the degree of variation of investment in different domains of expectations, for example, in a professional domain, different degrees of ambition. If I am more invested in my professional work, maybe I am less sensitive to collaborations and more to prestige, but the compensation ratio may be very different in another domain of expectations, for example relations between friends.

In order to complete the analogy (this is maybe the most disputable parts of the transposition), we need to find an analogue between the common inter-personal scale between several individual indifference curves between different goods, and, in the emotional intra-personal version, the different domains of expectations of the same individual.

The common inter-personal scale was given in the model of fair allocation by the total endowment of the relevant resources (bread and wine in the example). It might be given in our intra-personal analogy by the set of possibilities of satisfying our different expectations offered by our environment. In the example of the manager, this set includes the real possibilities—the capabilities—of social prestige, nice collaborations in the professional domain, as well as the possibilities of satisfaction in the domain of friendly relationships and the domain of autonomous creativity. Recurrent negative emotions in the domain of professional work, combined with more positive emotions in the friendship domain and in the creativity domain, show that resources of social prestige and nice collaborations are more scarce than expected, while the possibilities of friendship relations or creativity are richer and not entirely exploited. In this case the incentive to revise the relative priorities of work and of relations between friends or autonomous creativity is a rational one.

We may extend a bit further the analogy, as Fleurbaey and Maniquet try to come closer to more qualitative evaluations of real social situations by examining the possibility that some goods in a bundle are not cardinal and that more is not always better. They take the example of a bundle of goods in which the evaluation of one good can be cardinal and the other is only ordinal and exhibits satiation (they use the word "attribute" to include the two kinds of goods, as usually "goods" are likely to admit cardinal evaluation). In this situation, there is no common scale between the diverse indifferent curves, except if we can focus either on a worst attribute, or on a best attribute (a worst attribute is an attribute that is so painful that any bundle in which it is present is worst, whatever are the other attributes in the bundle, and mutatis mutandis for the best attribute). The authors suggest examining different combinations between rules of evaluation that were defined for cardinal and ordinal goods. Rules can use cardinal comparisons, or, in the other case, focus on a referent attributes. They can focus on the worst preferences, when the trade-off between the goods is difficult except for the worst goods that one has to avoid with highest priority, or on the best preferences, when all the goods can be compared.

Four combinations[4] are possible, and I will order them in a sequence of revisions, each revision being possibly triggered by an emotional revision. In the emotional analogue, goods have to be replaced by emotional aspects or values. They are not supposed to be reducible to cardinal evaluation, but for some of them, cardinal evaluations are acceptable proxys: financial situation, number of followers, number of those who agreed in opinion poll, etc. But it would be more sensible to replace "cardinality" by "inter-subjective stable ranking order", and "ordinal" by "subjective ranking".

(1) First combination: evaluating stable ranked values by giving priority to the best ranked, and the other values by focusing on the best referent attribute. We could easily find analogue emotional situations. This way of evaluation is similar to the one of a person that is attracted by an exciting perspective of gain (when the ranking of gains is stable), and is only sensitive to the top attributes in domains in which stable ordering is difficult. If she has not experienced the impact of recurrent negative emotions in one of this domains she may underestimate the worst attributes in this domain, and the risk of be deprived of a more basic but difficult to rank attribute. When she experiences such recurrent negative emotions, she becomes more prone to pay attention to worst attributes and to avoid to be deprived of more basic needs and values. Emotional revision (triggered by recurrent emotions) leads to combination 2.

(2) Evaluating stable ranked values by giving priority to worst ranked ones, and focusing for others values on the worst possible attribute. But sticking to worst things could lead to be depressed, and if she avoids breakdown, her recurrent emotion are incentives to shift to the third kind of evaluation.

---

[4] I have changed the order of the combinations in order to better fit the emotional dynamics.

(3) It would be better to focus on the value the deprivation of which would have been the worst situation for her but has also been fortunately prevented by the revision of her expectations. This revision may lead her to have a better and more stable evaluation of the relative ranks of her expectations in the relevant domain. In other domains in which no revision has brought such stabilizing effect, she still focus only on the best attribute. This move corresponds to combination 3: evaluating stable ranked values by the worst attribute and the other values by the best reference—but it is an unstable and insufficient combination, as the stability of these other domains has not been tested.
(4) Once she has moved back and forth and experiencing successively excitement, disappointment and relief in different domains, she has a better idea of what would be a sensible and more stabilized way of ordering her expectations across different domains: paying attention to values that she does not want to be deprived of, and using stable ranked values as references for making more encompassing comparisons. This is an analogue of combination 4: being able to evaluate best preferences for stable ranked values and focusing for the others on the worst attributes. I suggest that reaching this stability would not be justified —and is not really possible—without having experienced the whole emotional dynamics.

Conclusion

If we come back to the problem of temporal discounting, combination 4 seems also the way to avoid temporal inconsistencies. If I am tempted to stay at a party late in the night, but have tomorrow to pass an exam that is decisive for my career, the exam has surely a higher rank in the stable order of my values, and the anticipation of a failure in the exam is related to my worst perspectives. Nevertheless, even if emotional revision can lead to a more stabilized ordering of my domains of expectations and my values, this process takes a long time. Several bad experiences and failures are needed for recurrent emotions and their incentives to revise to bring forth their good effect. The only way for accelerating this process is to use the results of the emotional revisions of a previous generation for educating the following generations, in order for them to learn what are the worst attributes and what ranking order of values has been shown to be more stable.

One may believe that pure rational propositions can be discovered and learned more rapidly. This is true for learning, but maybe not for discovery. This is not true for what matters: our effective propensity to apply in practices in ordinary life the advices of rational knowledge. Rationality has a weak motivational power when confronted to affective motivations. Dynamics of rationality, dynamics of theoretical revision and discovery need the help of dynamics of emotional revision for being efficient.

References

Deonna, J. A., & Teroni, F. (2012). *The Emotions, a philosophical introduction*. London and New York: Routledge.

Fleurbaey, M., & Maniquet, F. (2012). *Equality of opportunity: the economics of responsability*.

Frederick, S., Loewenstein, G., & O'Donoghue, T. (2002). Time discounting and time preference: A critical review. *Journal of Economic Literature, 40,* 351–401.

Frijda, N. (1986). *The Emotions*. Cambridge: Cambridge University Press.

Geoffard, P. Y., & Luchini, S., "Changing time and emotions". (2010) In A. Kirman, P. Livet & M. Teschl (Eds.), *"Rationality and emotions"*, *Philosophical Transactions of the Royal Society B*, Vol. 365, n 1538, pp. 271–280.

Kahneman, D., & Tversky, A. (1984). Choice, values and frames. *American Psychologist, 39*(4), 341–350.

Kahneman, D., & Tversky, A., (2000). *Choice, values and frames*. In Kahneman & Tversky (Eds.), Cambridge: Mass. Cambridge University Press.

Han, R., & Takahashi, T. (2012). Psychophysics of time perception and valuation in temporal discounting. *Physica A, 391,* 6568–6576.

Livet, P. (2002). *Emotions et rationalité morale*. Paris: PUF.

Livet, P. (2010). Rational choice, neuroeconomics and mixed emotions. In A. Kirman, P. Livet & M. Teschl (Eds.), *"Rationality and emotions"*, *Philosophical transactions of the Royal Society B*, Vol. 365, n 1538, pp. 259–269.

Livet, P. (2016). "Emotions, Beliefs and Revisions", *Emotion Review*.

Livet, P. (2017). *OEconomia*, (7–2), 191–200.

Rott, H. (2001). *Change, choice and inference: A study of belief and revision and nonmonotonic reasoning*. New York: Oxford University Press.

Rott, H. (2002). « Logique et choix », ch. 5 in *Révision des croyances*. P. Livet (Ed.), Paris : Hermès Lavoisier.

Schlechta, K. (1997). *Non monotonic logics: Basic concepts, results and techniques*. Heidelberg, Germany: Springer.

Schlechta, K. (2004). *Coherent systems*. Heidelberg, Germany: Springer.

Chapter 12
Pathologizing Ideology, Epistemic Modesty and Instrumental Rationality

Leslie Marsh

Abstract Practical politics is plagued by an unabashed and unrelenting mutual demonization of a given ideology, with each side classing each other's cluster of ideas as pathological, i.e. "indicative of disease" and/or "extreme or excessive" or hopelessly irrational. This chapter specifically argues that the demonization, or the pathologization of conservatism as an ideology runs on a straw man fallacy—that is, detractors (*and* vulgar catechumen-like defenders) blithely assume that conservatism is coextensive with an ideology. This chapter argues for the view that not only is political conservatism *not* an ideological worldview, it is a cluster of *epistemic virtues* that should temper the rationalistic impulse *regardless of ideological commitments*—at least within the domain of sociality. Epistemic conservatism in its most generic form is the idea that a belief has some presumption of rationality merely because it is held. Cognitive closure, otherwise known pejoratively as "new mysterianism", is the view that the mind is structurally constrained in its computational power. Situated cognition, or ecological rationality, is a stance emphasizing rationality as being constitutive of activity, context, and culture. Social externalism is the view that much of our thinking is individuated in part by the linguistic and social practices of a thinker's community. The social complexity thesis is the view that there cannot be a predictive science of politics to drive a radical reconstitution of society. Complexity in the social realm, intrinsically stochastic, is coordinated by a voluntary manifold of self-organizing emergent spontaneous orders. Though each of these theses have resonance to so-called "political" conservatism they are not political positions per se. Regardless of ideological commitments, a cast of mind displaying a significant over-preponderance of rationalist traits, could well be deemed a neurodevelopmental disorder.

L. Marsh (✉)
The University of British Columbia, Vancouver, Canada
e-mail: leslie.marsh@ubc.ca

© Springer International Publishing AG, part of Springer Nature 2018   165
G. Bronner and F. Di Iorio (eds.), *The Mystery of Rationality*,
https://doi.org/10.1007/978-3-319-94028-1_12

Preamble

Epistemic conservatism in its most generic form is the idea that a belief has some presumption of rationality *merely* because it is held (Quine). Cognitive closure, otherwise known pejoratively as "new mysterianism", is the view that the mind is structurally constrained in its computational power (Hayek, McGinn, Simon). Situated cognition, or ecological rationality, is a stance emphasizing rationality as being constitutive of activity, context, and culture (Hayek, Polanyi, Oakeshott, Vernon Smith, Gigerenzer). Social externalism is the view that much of our thinking is individuated in part by the linguistic and social practices of a thinker's community (Wittgenstein, Davidson, Burge). The social complexity thesis is the view that there cannot be a predictive science of politics to drive a radical reconstitution of society (Popper, Oakeshott, Hayek). Complexity in the social realm, intrinsically stochastic, is coordinated by a voluntary manifold of self-organizing emergent spontaneous orders (Adam Smith, Ferguson, Grassé, Hayek). Though all these theses would have resonance to so-called *political* "conservatism" *they are not* political positions per se. If there is one collecting feature that gathers this group of ideas it is their anti-rationalism. Therein lies the rub. For reasons I will explain, not only is political conservatism not an ideological worldview, it is a cluster of *epistemic* virtues that should temper the rationalistic impulse regardless of ideological commitments, at least within the domain of sociality.

Practical politics is plagued by an unabashed and unrelenting mutual demonization of a given ideology, with each side classing each other's cluster of ideas as pathological.[1] By "pathologizing", the intent is unmistakenly derogatory, connoting both meanings of the word—"indicative of disease" and "extreme or excessive". As such, the other is deemed *illegitimate* and *irrational*. The upshot is that the broad centre, that is, the space for reasonable[2] disagreement, *the* most vital constituent of the civil-liberal condition, has been severely corroded. Most of the vitriol has been targeted at conservatism, an "academic"[3] vitriol often indistinguishable from

---

[1]Schmitt (1927/1976) famously took the view that integral to politics is a friend-foe relationship, a relation holding between groups not individuals. Corey (2014) coined the neologism "dogmatomachy" from the Greek *dogma* (an opinion that falls short of knowledge) and *machē* (battle). Others take a *forum* view of politics, one that is constrained by obviously political institutions— e.g. the British parliament.

[2]By "reasonable" one doesn't mean to imply that free-speech should in *any* way be curtailed.

[3]onservative theorizing is caught between a rock and a hard place. Analytical philosophy, in line with much of academia, resents this primary colour, something that should necessarily be part and parcel of a balanced epistemological pallet: Duarte et al. (2015) note that the lack of *viewpoint* diversity conspicuously missing from social psychology circles, must surely have epistemic consequences for psychology. Many a self-identified conservative theorist in turn has resisted an analytical approach (Beckstein and Cheneval 2016). Sensitive to the supposed anti-intellectualism attributed to conservatism, Roger Scruton notes that the grand rationalistic systemization of political theorizing inherently confers a smug and self-congratulatory attitude upon proponents (see note 4); this in contrast to the conservative emphasis in articulating "the reasons for not having reasons"

tabloid-style outrage, hurling tired, virtue-signaling[4] epithets such as reactionary, monarchist, papalist, ecclesiastical, agrarian, aristocratic, feudal, fascist, and so on.[5] This chapter argues that the demonization, or the pathologization of conservatism as an ideology runs on a straw man fallacy—that is, detractors blithely assume that conservatism is coextensive with an ideology that in turn supposedly appeals to a personality type on the authoritarian spectrum.[6] The Left-Right spectrum has long since lost its conceptual value (Abel and Marsh 2014): a richer dimension is authoritarian-liberal,[7] which has the twofold consequence that:

(a) Left and Right extremism are species of one cast of mind—a rationalistic one; and that

(b) conservatism, properly speaking, is very much a liberal position.

Things are, of course, muddied on the grounds that self-avowed conservatives are complicitous in promoting this fallacy by suggesting that there really is a distinctive matching ideology. That conservatism as an epistemic stance has no traction in practical politics is not surprising: supposed conservative thinking has been bastardized for easy consumption. "Conservatism" as Kristóf Nyíri observes, "is not necessarily right-wing" (Nyíri 2016, p. 445). As an epistemic stance, all that *political* conservatism claims is that we do not have a predictive science of politics on grounds of *complexity*; and that it is epistemically prudential, for a whole tissue of reasons, to

---

[4]A platitudinous gesture to "self-brand" oneself, a form of vanity, dressed up as selfless conviction: "people are trying harder to *look* right than be right" (Haidt 2012, p. 89; emphasis added). Taleb's scathing polemic articulates the generalized psychological driver behind this impulse: "The IYI [The Intellectual Yet Idiot] pathologizes others for doing things he doesn't understand without ever realizing it is *his* understanding that may be limited" (Taleb 2018, p. 124). As Susan Haack puts it, this is no more than "activism masquerading as inquiry" (Haack 1998), the *sine qua non* of the clerisy that is characteristic of the prevailing "professional" thinker.

[5]This lack of conceptual precision is tied to what has been termed as "conceptual creep" (Haslam 2016; Haidt 2016). Indeed, Percy long since pointed this out: "Nothing much is proved except that current categories and names, liberal and conservative, are weary past all thinking of it. Ideological words have a way of wearing thin and then, having lost their meanings, being used like switch-blades against the enemy of the moment". Percy (1991, pp. 58; 416). Percy pointedly wrote (1991, p. 92) out that "Political conservatism is neither sinful nor illegal—though sometimes one wonders if liberals don't think it is".

[6]Academic social theorizing, now operating primarily under the shadow of Marxist "false consciousness" (or *mis*cognition) and wedded to radical social constructivism (often deeply at odds with the prevailing best science; Marsh 2005; Nyíri 2016), has been the major driver behind the significant narrowing of the Overton window. Duarte et al. (2015) note that the lack of *viewpoint* diversity conspicuously missing from social psychology circles, must surely have epistemic consequences for psychology. Most notably has been the consistent failure of Implicit Association Testing in passing standard tests of replicability (Lilienfeld 2017).

As Susan Haack puts it, "activism masquerading as inquiry" (Haack 1998) is the bread and butter of the priestly class of "professional" thinker.

[7]Mathematician Eric Weinstein's (2016) Four Quadrant Model similarly illustrates how the media stigmatizes certain nuanced views that challenge the status quo by portraying people who hold those views as prejudiced or intolerant. For Percy (1971) ideologues are the bore crashing the conversation of mankind.

preserve the existing, albeit flawed, advantages, rather than to instigate a wholesale trading in of inherited practices for the completely unknown. Whatever the limitations, contradictions, paradoxes, incoherencies, anomalies, and other flaws in a given tradition, one always has to work from within that tradition. To do otherwise would not only be irrational but would also be reckless.[8] Both self-avowed political conservatives and their critics are blind to the contention presented here that conservatism is a prudential epistemic stance. *While not domain-specific, it does provide particularly vital insight into maintaining a dynamic liberal order.* In other words, conservatism plays a vital epistemic foil in assessing the claims and limitations of what is variously called means-end rationality, instrumental rationality or rationalism.

Politics at the best of times is a defective experience. No country has the resources to satisfy everything claimed as a human right by its *cives*. The extension of- and/or the reduction of all to- the political, renders not only practical reasoning epistemologically suspect, but it also tarnishes other domains of experience that have been subsumed, or coopted, by politics.[9]

The perfectability of human nature lies at the core of the rationalists' mistaken view of human reality—i.e., prohibition (Nyíri 2016). Tone-deaf to the myth of Original Sin and, for that matter, the myth of the Tower of Babel, "[t]he rationalist imagines an imbecile-free society; the empiricist an imbecile-proof one, or, even better, a rationalist-proof one" (Taleb 2016, p. 100). In other words, *the rationalistic cast of mind is the identifiably pathological condition, regardless of the ideological constellation.* It is, to reuse Wittgenstein's famous phrase, a bewitchment of our intelligence. I seek to dispel the view that ideological constellations are inevitably pathological (despite legitimate ascriptions of scientism, collectivism, fundamentalism, atomism, and so on), but *only insofar as there is an overemphasis on the rationalistic component.* "Perfect" knowledge is unnecessary and impracticable. Most worryingly, its relentless pursuit inevitably generates *il*liberal outcomes.

The discussion proceeds as follows. In the next section, by way of some ground clearing I present a conceptual analysis of liberalism and conservatism, examining their continuities and discontinuities. In § 3, I present the salient outlines of Hume's moral psychology. In § 4 I discuss the recent work of two prominent cognitive scientists who have both analyzed the "conservative mindset" though to very different purpose. In § 5 I critically discuss Michael Oakeshott's famous "dispositional conservatism", tying him directly to the Humean tradition. Finally, I return to the opening Quinean gambit regarding epistemic conservatism.

---

[8]A case in point being Angela Merkel's display of hubris syndrome traits with regard to the migrant crisis: "a particular form of incompetence when impulsivity, recklessness and frequent inattention to detail predominate. This can result in disastrous leadership and cause damage on a large scale. The attendant loss of capacity to make rational decisions is perceived by the general public to be more than 'just making a mistake'" (Owen and Davidson 2009; see also Murray 2017).

[9]The prevailing view in the academy is that power and politics are in lockstep (Michel Foucault), thereby making politics a ubiquitous activity (See Robin 2011 as an preeminent instance).

Ideological Morphology

The hardest-won and greatest achievement of the liberal tradition (as will be explained, this includes conservatism) has been the epistemic independence wrestled from concentrations of power, monopolies and capricious zealotries (Hardwick and Marsh 2012a, b). Whatever else might be attributed to liberalism, it has primarily embodied the idea that conceptions of the good and goals of action are irreducibly plural. Indeed, the very precondition of knowledge is a generalized exploitation of the epistemic virtues accorded by liberal society's *distributed* manifold of spontaneous orders and forms of life, giving context and definition to intimate, regulate, and inform action (Boettke 2017; Haack 2012).

Though liberalism has always had a strong rationalistic component (i.e. the proto-liberalism of Hobbes), the 20th Century saw a vast expansion of (top-down) rationalism at the expense of the ever-dynamic and fine-grained complexity that is sociality. The primary instrument of salvation effecting this social engineering has been the state, a monopolistic, blunt and coercive mechanism. The arguably necessary and inherent tension between a conception of the state as non-instrumental versus one as instrumental in the promotion of some substantive theory of the good is so out of equilibrium that we are now witnessing a profound fracture within the political class, their constituents, and, consequently a deep corrosion of the broad centre vital to the maintenance of a genuinely liberal culture.

Liberalism or more accurately liberalism*s*, *conservatism included* (Alexander 2015, p. 10, 2016, p. 14), do not constitute a homogenous and internally consistent body of ideas. There are, broadly, three variants of liberalism believing respectively that: (a) there is known to be an objective human good, at least its nature is a matter of reasonable belief (Locke's "reasonableness of Christianity"); (b) there is an objective human good but our knowledge of it is incomplete (Mill's "experiments of living"); and (c) there is no such thing as an objective good: a person's interests are defined in terms of their preferences, desires and inclinations (the hallmark of market capitalism).

The *typical*[10] features of liberalism are:

- individualism (individuals are the ultimate units of moral value/society has as its proper end the good of individuals/individual well-being requires people to make their own choices as far as possible)
- universalism (affirmation of the moral unity of the human species, according secondary importance to historical circumstance)

---

[10]As with all ideological analyses, none can be tidily encased in a set of necessary and sufficient conditions. Different ideologies cluster concepts differently; some concepts are sidelined, downplayed, emphasized, or reinterpreted. So, for example, conservatives do not conceive "social" justice to be purely a matter of distributive justice as is also the case for libertarians, but for somewhat different reasons. Libertarianism, derivative of liberalism, has over-sacralized one value to the detriment of all others—in this case, the market (see Abel and Marsh 2014). See Iyer et al. (2012) on the Libertarian disposition.

- egalitarianism (all mankind has the same moral status)
- meliorism (the affirmation of the corrigibility and improvability of social institutions and political arrangements).

It should be noted that conservative constellation does embody the above features though each component is differently weighted and articulated with an emphasis on:

- scepticism (anti-rationalism and the complexity thesis)
- the rejection of the politics of ideals (limited, role-based government)
- reliance on practical reasoning by which a tradition can be interrogated and applied (one is always dealing with a reflective tradition, *not* an inert pattern of habitual behavior) (Beckstein 2017)
- organicism and communitarianism (stressing the situatedness of the self and the rejection of focal/abstract individualism).

Six qualifications are called for. First, conservatism can be historically specific—e.g. Burke, de Maistre, and, to a degree, Aristotle and Hume. Second, it can be indexical and it is this notion that tends to confuse its detractors. The idea is that one can be conservative relative to a context, while in no way holding any recognizable conservative values; for example, anyone who wished to preserve the essentials of Communist Party rule was deemed a conservative. Third, liberalism and conservatism (and socialism for that matter) should not be conceived as being coextensive with self-identifying political parties or movements; this is not a reliable indicator of their ideological commitments and this further is muddied by trans-Atlantic terminology (Marquez 2015, p. 410). Fourth, the blithe conjoining of terms such as "liberal" to "democracy" is a commonplace misnomer: liberalism does *not* entail democracy. Fifth, it's worth noting that being a conservative in politics does not necessitate being a social conservative: there is only a contingent relation. Lastly, while value incommensurability emphasizes choices between freedoms they are neither mutually entailing conceptually or practically consistent.[11]

The point of contention revolves around the melioristic aspect of liberalism and the sceptical aspect of conservatism. The prevailing "liberalism" has over-emphasized its commitment to meliorism, ditching the notion of individualism for the anti-individualistic bloated social ontologies characteristic of identity politics.[12] Though meliorism is an important strand in the liberal ideological constellation, one should be very cautious in aligning liberalism exclusively with progressivism. Liberalism (at least in its classical variety) is *not* coextensive with progressivism.

---

[11] See Eyal and Tieffenbach (2016) on incommensurabilty and (market) value.

[12] It is this outlook that has, not surprisingly, tilled the soil for, what in current jargon is known as "intersectionality", an ever-expanding, voracious hierarchical basket of rights-claims running on irrelevant collecting features making for an inherently divisive political culture. In identity politics the *explanans* (a particular statement, law, theory or fact) and the thing to be explained (the *explanandum*) have been inverted (so, for example, it would be deemed irrational for non-conformists holding feature x not to subscribe to an expected ideology despite their superficially similar characteristics), e.g. self-identifying Black or gay conservatives.

Traditional liberalism is concerned with equality of *opportunity* (equality) and not equality of *outcome* (equity) as is the vain impulse of the progressive rationalist. The former understands that human behavior is stochastic and therefore to insist on equity of outcome requires relentless overreach that must thereby embody a distinctly authoritarian flavor, if that were even minimally practicable.

Epistemic humility is not seen as a cultural virtue: it is the *zeitgeist* of the current age that we exist in a (misperceived) linear trajectory of progress, progress here taken to be coextensive with improvement—morally, socially, technologically, economically and scientifically. Progressivism is clearly a "grand narrative" notion which on closer scrutiny is subject to all the weaknesses of such constructions. It is impossible to determine whether a change for the better in one part or aspect of the system is progressive for the system overall, since there is no Archimedean point from which progress can be assessed. Every change alters some state of affairs, destroying or modifying it. Granted we live, in some real sense, the best of times (reductions in child mortality, vaccine -preventable diseases, access to safe water and sanitation, malaria prevention and control, prevention and control of HIV/AIDS, tuberculosis control and declining poverty—Pinker 2018). We also live in the worst of times—Auschwitz-Birkenau, Holodomor, Cambodia and more besides —the dark side to political technocracy.

Conservatism has taken up liberalism's original emphasis on scepticism, as per the independence wrought from epistemic monopolies mentioned at the beginning of this section. So, in this sense, conservatism is very much part and parcel of the liberal tradition (or as Irving Kristol so memorably quipped "a liberal who has been mugged by reality"): and it is this that seems to be overlooked by not only its detractors but quite often by self-ascribed conservatives themselves (Podoksik 2008).

The vital point is that ideologies are porous and as such are morphological: that is, there is a great deal of fluidity within and between ideologies and thinkers can and do pass through one ideology to another (Freeden 1994; O'Sullivan 1976; Alexander 2015). On this account notions of ideological purity must be laid to rest. So for example, the communitarian element in conservatism (*and* socialism; Alexander 2015, pp. 14–15, 17, 20) does not make communitarianism a conservative perspective per se. Neither can communitarians, conservatives or liberals accept the market as the *dominant* model for social relationships, as radical libertarians are wont to do—which is not to say that they are in any way anti-market (Hardwick and Marsh 2012a). This ideological porousness is manifest in a Darwinian Leftism proposed by Singer (2000, pp. 60–63). Singer's checklist requires that any Leftism worthy of its name cannot: (a) deny the existence of human nature, or insist that human nature is inherently good, or that it is infinitely malleable; (b) expect to end all conflict and strife whether by political revolution, social change, or better education; (c) assume that all inequalities are due to discrimination, prejudice, oppression, or social conditioning. Some will be, but this cannot be assumed in every case; (d) expect that, under different social and economic systems, many people will act competitively in order to enhance their own status, gain a position of power, and/or advance their interests and those of their kin; (e) stand by the traditional values of the Left by being on the side of the weak, poor and oppressed, but think very carefully about what social and economic changes will really work to benefit them.

Items (a) through to (e) are items that the conservative can and should sign up to. In other words, these are the central values of our shared inherited liberal culture. The problem is that, regarding the "regressive Left's"[13] ostensibly most cherished value, (item e)[14] only gives lip service via meaningless social ontologies, generating a divisive "new" racism, that is, identity politics with "rights talk" as its handmaiden.

The idea of ideological morphology has serious consequences for the patholo-gizing view that there are recognizably impervious ideological identities, and that there must therefore be some mileage to the associated notion "polylogism". Polylogism, attributed to (von Mises 1949) is used here to connote no more than the idea that epistemological outlooks differ between human groups. It is also rela-tivistic in the sense that it's a weak or trivial type of relativism tied to a form of life, and does not express any absolute cross-cultural truths (Williams 1985).

A Hume Primer

As indicated at the outset, the thesis presented here is that Hume's philosophical psychology provides the most compelling understanding of sociality and its asso-ciated situated rationalities. The overarching significance of Hume to the discussion is that for Hume being reasonable meant accepting the limits of reason, a sentiment so crisply expressed by mathematician Ken Benmore: "There is ... nothing irra-tional about a philosopher who isn't a rationalist" (Binmore 2011, p. 11). Alert to the socio-political situation of Hume's day, one must be careful to not simplistically refract Hume's outlook through the post-Revolutionary bifocal liberal-conservative lens as represented by the triumvirate of Hume, Burke and Smith. As David Miller puts it, "Hume believed that those things which liberals characteristically value are indeed valuable, *provided* that those things which conservatives characteristically value can be securely enjoyed at the same time" (Miller 1981, p. 195).

Whether one takes the view that Hume is a grandee of the representational theory of the mind tradition (Garrett 2006; Landy 2012), or whether one thinks that Hume plays the pivotal role to the non-Cartesian *situated* tradition (as I and Froese 2009 do) the *Treatise* (Hume 1978) is "the foundational document of cognitive science" (Fodor 2003, p. 135). Though Hume did not have a modern scientific toolbox at his

---

[13]The term "regressive Left" was coined by Nawaz (2016, pp. 210, 251). The policing of language or compelled speech language being a star example of a stochastic spontaneous order, necessarily invites a perpetually authoritarian response in enforcement because of the quicksilver nature of complex systems; a notable slide towards a Koestler-like *Sonnenfinsternis* moment has now been set in motion by the Canadian government's adoption of Motion-103 and Bill C-16.

[14]Looking after those less fortunate can also be found as a prime social value in idea of "noblesse oblige" of High Toryism.

disposal, I take "sociobiology" (called "evolutionary psychology" these days) to be a scientific vindication of Hume's speculative anthropology.[15] Hume's moral psychology has, of late, found new voice within the recent trend that has come to be known as experimental philosophy, most notably in the work of Jonathan Haidt.

Hume's *Treatise*'s full subtitle is *Being an Attempt to Introduce the Experimental Method of Reasoning into Moral Subjects*. By "experimental" Hume means "based on knowledge and observation", taking due account of the relevant differences between the human and physical sciences (*Treatise* xix). For Hume (1975) moral judgment occurs when one (a) checks all the relevant facts (*Enquiry* 173) and (b) takes a general view. By the former, Hume means that, within practical cognitive limits salient facts inform the utility or agreeableness of a character trait. This is what we'd call being "well-informed". By the latter, Hume means, that in moral assessing someone's action, one must be disinterested, a detachment understood as sympathy (*Treatise* 472, 579). Morality, says Hume, "is more properly felt than judged of" (*Treatise* 470).

For Hume, moral judgments are causal through emotion and not through logical implication. Hume takes the view that agents have a direct inclination to do certain actions because there are corresponding character traits that comprise human nature. For Hume, a *natural* virtue is a morally good *tendency* that can be exercised independently of a social context (though within limits, they do appear even as though they are). Moreover, in the exercise of these traits, we regard the interests of others as essentially and intrinsically (not accidentally or instrumentally) to one's own advantage. The limits alluded to are motivated by self-regarding and limited generosity actions. Generosity is limited by the degree of sympathy, which, in turn, is limited by a degree of propinquity. This propinquity, mediated by social institutions, radiates through loyalties and allegiances and that while actions may attract moral approbation or disapproval, are not built into one's character traits. While the natural virtues delimit the bounds of generosity, the *artificial virtues* are grounded in self-interest. For example, justice (an artificial virtue) serves one's self-interest in maintaining the institution of property, from which we all benefit. Hume does not mean that these artificial virtues are somehow spurious, merely that they are contingent upon the contrivances of sociality. Moreover, they are a part of morality where certain general conditions of trust obtain.

A standard criticism leveled at Hume is that his argument relies too heavily on the independent spectator's viewpoint rather than on the agent's viewpoint (more on this in the next section). If there is a vulnerability, it lies within Hume's emotivist approach: the idea that moral judgment does not embody any genuine

---

[15]E. O. Wilson's "sociobiology" of the mid-70 s was deemed too controversial, so much so that he was, sadly, viewed pretty much as a pariah for two decades afterward. As Haidt (2013b, p. 282) more elegantly puts it: "I got the feeling that sociobiology was radioactive. It was dismissed as reductionist and it was tainted as a gateway theory leading to racism and sexism." The regressive Left's anti-science bent has now been fully unmasked by ex-Google employee, James Demore's memo: https://assets.documentcloud.org/documents/3914586/Googles-Ideological-Echo-Chamber.pdf.

knowledge of an independent reality.[16] For Hume, approbation and disapprobation are non-reducible experiences, much like a specific qualic experience. So what of the *objects* of moral judgment? For Hume this means that one's feelings of approbation and disapprobation are expressive of *character*. Character, in Hume's associationist psychology, is that which is "durable enough to affect our sentiments" (*Treatise* 575). The upshot of Hume's position is to cordon off the affective (perhaps impervious) from reason. Human behavior thus conceived is governed primarily by unanalyzed experience or habits. Hume's slogan "custom is the great guide of human life" is quite compatible with his near contemporary Bishop Butler's quote "probability is the very guide of life".

All this points to a deeper aspect of Hume's argument. Hume's intention was to undermine the very idea of an underlying rational harmony in nature, the *sine qua non* of 18th Century rationalism. Hume assaulted this assumption by contending that the external order is not merely a product a priori reasoning but that its discovery was rooted in the principles of human nature. The phenomenal world rests on conviction and not on a process of logical inference. This marks the dispute between the necessatarians and the regulatarians; Hume's position famously the latter. Hume's idea is this. No particular analysis of the concept of cause commits one to the principle of causality—i.e. that every event has a cause. All we actually experience is that one kind of event is constantly conjoined with another. A science of politics, therefore, must be grounded in experience supplied by historical inquiry and observation of existing societies. Essentially, it was to be an investigation into the interaction between institutions and human nature. In this way Hume indicated to later conservatives (and Leftists such as Singer) that the strongest arguments for the existing order were to be found within the facts of that order; that under an empirical approach, utility could be located as an immanent value embodied in *actual* social arrangements, rather than in the tendency to exaggerate principles into unyielding absolutes. The latter, of course, is a mark of the rationalist across the ideological spectrum and in no uncertain terms should be deemed a pathological condition. This said, one cannot really ascribe a specifically liberal or conservative label to Hume. In the *Essays*, Hume (1985) rejected the social contract liberalism of Locke, but was somewhat too sceptical a thinker for Burke—both of whom could be said to plausibly straddle both these labels. "Hume was something more than the Enlightenment incarnate, for his significance is that he turned against the Enlightenment its own weapons" (Wolin 1954, p. 1001). McArthur argues that if Hume is a conservative, he is not of the traditionalist variety, but more of the precautionary variety; both views can be simultaneously coherently held (McArthur 2007, p. 123).

---

[16]A. J. Ayer and C. L. Stevenson offer the two classic twentieth century formulations.

Pathologizing Ideology

One of the most prominent figures in promoting the pathologization of conservatism is George Lakoff.[17] Lakoff's slogan that "all politics is moral" echoes the commonly made claim that political philosophy is simply the application of moral philosophy to public affairs. Lakoff is well-known within non-Cartesian situated cognition circles. As such he rejects the Enlightenment's conception of abstract unvarnished universal rationality, at least concerning the paradigmatic complexity of social matters (Lakoff 2008, pp. 1–3, 6–10). Lakoff clearly is in accord with Hume in that he takes the view that most reason is subconscious. So why would he feel the need to pathologize conservatism, especially since conservatism would be in full accord with Lakoff's critique of the rationalistic aspects of Enlightenment thinking? The reason lies in Lakoff's role as an activist and polemicist. He claims that in the so-called "culture" war progressives "have ceded the political mind to conservatives" (Lakoff 2008, p. 2).[18]

First, Lakoff's use of the term "radical conservative" is literally a contradiction in terms (Lakoff 2008, pp. 42, 43, 68). If he's referring to so-called "religious fundamentalism", then that is already disqualified on conservative terms because of the deeply rationalistic impulse that drives this fundamentalism. Second, Lakoff's linkage of conservatism to *laissez-fairism* (Lakoff 2008, p. 62) is equally misleading since a spontaneous order (as a market is) supposedly corrodes traditional patterns of behavior. While there is something to this claim, it's a view I don't happen to share. For if traditional patterns of behavior are not classic instantiations of spontaneous order, then what are they? Third, the idea that the political divide is coextensive with a fundamentally democratic *normalized* (liberal) and a fundamentally antidemocratic *pathologized* (conservative) ideology simply does not hold (Lakoff 2008, p. 5)—recall my admonition that liberalism (and derivatively conservatism) do not entail democracy. Antidemocratic sentiment has a tradition going back to Plato and can be found in radical versions across the political divide.

On Lakoff's view, progressivism is akin to the "nurturing" parent, which accords with what Haidt has identified as the progressive's primary moral value: care. Correspondingly, the other parent is the "authoritarian father" figure, which Lakoff

---

[17]Others include Honderich (1991); Jost et al. (2003) make the feigned sounding claim to the contrary: "This does not mean that conservatism is pathological or that conservative beliefs are necessarily false, irrational, or unprincipled" (Jost et al. 2003, p. 340). The latest addition to this genre of crude polemics is Robin (2011). It is telling that for Robin power, à la Lenin, is *the* central primary question without ever making any distinctions as per (Dahl 1957) or Russell (1938). Moreover, Robin overly conflates power with authority, and in turn, authority with hierarchy.

[18]This claim is odd since in the 30 years that I've been monitoring the so-called "culture wars", the right has been operating within the wake of the Left's hegemony, especially in education. Though it was the Right of the '80s and '90s that harboured anti-science (debates about evolution) and censorious tendencies (music labeling, actually in cahoots with the left), this stance is now more applicable to the "regressive" Left). Lakoff's book comes over as having an ideological to axe to grind and as such is weakened with infelicities.

ascribes to the conservative mindset. Are these two casts of mind mutually exclusive? (cf. Kanai et al. 2011; Kaplan et al. 2016). Not according to Lakoff whose "biconceptualism" accepts that many retain both conservative and progressive views, which are made manifest in different contexts and on different issues. As I've already indicated, this points to the fluidity of ideological clusters. It might be argued that if socio-political rationality is infused by some supposed, even minimal, knowledge of the conditions for human flourishing, then care of all *cives* should be any ideology's central concern. When Lakoff says that since political positions are neither logical nor self-made, they can and should be altered, and that facts or propositional thinking is not required for persuasion, this perhaps explains the postmodern veriphobic climate under which we now lived for nigh on 50 years.

Despite being aware of the elitist connotations of (by definition) top-down rationalism, Lakoff is clear in his view that "the role of the progressive government is to maximize our freedom" (Lakoff 2008, p. 48). We've already marked the problem with the notion of "progress" and that the instrument to effect this can only be the state. This makes positive freedom the primary political value, the idea that an agent's own character or personality can work against the agent's own interests. Positive freedom points away from the divided self, but this unity comes at the cost of privileging a favored political value. The risk a dominant positive freedom theorist runs is that this idea needs to be conjoined with negative freedom if some notion of autonomy is to be preserved. Lakoff pretty much denigrates the conservative mind set as essentially defective, *despite* the his recommending that the progressive constituency could learn much from conservatives.

Many along the liberal-conservative axis are in full accord with Lakoff's dissatisfaction with the Enlightenment[19]; that is, the privileging of propositional rationality be it of the collective planning variant or the abstract fiction of *homo economicus*. Where they part company with him is that the Continental Enlightenment was not the sole Enlightenment—the other tradition being of course the Scottish Enlightenment of Hume, Smith and several others. Though both traditions trumpeted the sovereignty of human reason, they differed on the legitimate scope of its application. The French variant put store in centralized top-down governmental coercion. By contrast, the Scottish variant took the methodological view that since reason is a property of an actual individually situated mind, it is neither practicable nor indeed desirable to aggregate the multitudinous individual hopes, dreams and aspirations. The Scots placed their discussions about moral philosophy within the framework of an intellectual legacy inherited from seventeenth—century European discussions about morality and natural law: man was, by nature, a social animal and the social world was defined by a complex network of authority and mutual obligation to one's fellow citizens.

Politics, for Lakoff is *not* about changing minds by deploying primarily a rational argumentative strategy. Politics is more about propagating *emotionally*

---

[19]A far deeper critique can be found in the work of Alasdair MacIntyre.

charged narratives and metaphors. He believes conservatives have been eminently successful in this—as does Jonathan Haidt.[20]

The Haidt Report

While George Lakoff is an unapologetic progressive activist, his research appearing incidental to underpinning his political polemics, Jonathan Haidt by contrast, genuinely wants "to understand moral disagreement and help people *disagree well*" (Perry 2016, p. 70, emphasis added). Haidt is through and through a Humean. In the service of this, he offers a closer-grained analysis of why conservatism has been so effective. According to Haidt, conservatism (or if as I prefer, *embedded* or *situated liberalism*) appeals in even measure *across* the six drivers of moral psychology (noting their corollary)—**care**/harm, **liberty**/oppression, **fairness**/cheating, **loyalty**/betrayal, **authority**/subversion, **sanctity**/degradation. This is the core of Haidt's moral foundations theory (2009, 2012). What Haidt has noticed across his studies is that the so-called liberal drivers are very much narrower in *privileging* the three pillars care/liberty/fairness and *deemphasizing* the pillars of loyalty/authority/sanctity.[21,22] The upshot of this is that conservatives' broader moral matrix allows them to detect threats to moral capital that "liberals" cannot perceive (Haidt 2008, pp. 343, 357; Amodio et al. 2007; Hibbing et al. 2014; Talhelm et al. 2015; Verhulst et al. 2012, 2016; Zamboni et al. 2009). To put it in more colloquial terms, conservatives really do have a wider moral bandwidth. Hence it should be no surprise that Hume is often held up as the first modern conservative thinker. Moral psychology for Haidt (and of course Hume) is the key to understanding humanity (Haidt 2008, p. 231) and in turn politics. Put another way, those on the Right have understood that the elephant, not the rider, is in charge of political behavior (Haidt 2008, p. 181)—this, as we've seen, is acknowledged by Lakoff. The moral vision offered is constrained by the progressivists' fetishizing "victimhood" (Haidt 2012, p. 182), an ever complicated permutation hierarchy of identities, or "the True Taste restaurant, serving up a one-receptor morality" (Haidt 2012, p. 141) whereas conservatives' morality appeals to all five moral "taste receptors"/moral matrixes.

---

[20]Haidt anecdotally mentions that as a self-identifying liberal he had tried to offer the Democrats advice on how to beat Republicans at their own game—the offer was not taken up.

[21]Burton (2015) tentatively point to the higher degree of neuroticism amongst "liberals" compared with "conservatives".

[22]A critic of Haidt suggests that "fairness and relief of suffering are more fundamental values than authority and loyalty, which are virtues only if their objects are worthy" (Blum 2013). In other words, Blum is saying that blind loyalty to an authority is unacceptable. Blind loyalty is not a feature of writers even as early as Burke. Blum would need to spell out with far more conceptual discrimination the varieties of- and the logic of- loyalty, *de jure* authority generalized as follows: A has authority over B if and only if the fact that A requires B to φ (i) gives B a content-independent reason to φ and (ii) excludes some of B's reason for not-φ-ing (Green 1988).

Since the "True Taste restaurant" offers a mono-metric scale, it is inevitable that "ideological purity" tests are bound to plague the progressivists' own worldview.

Haidt's diagnosis is that the Left in its over sacralization of equality to include equality of *outcome* is bound to be frustrated by the spontaneous order that is the market, culture, language and sociality and, most importantly, each individual's unreflective rationality (Haidt 2012, p. 204). This overreach is akin to a glacier relentlessly grinding and flattening an ever trivial social landscape—and this inevitably carries authoritarian tendencies to effect policy. The rationalistic mind falls in love with its own creations. A purely rationalistic approach to politics is bound to perpetually frustrate as a "mass delusion" (Haidt 2013a, or as an ideological bed of Procrustes:

> Since the Enlightenment, in the great tension between rationalism (how we would like things to be so they make sense to us) and empiricism (how things are), we have been blaming the world for not fitting the beds of "rational" models, have tried to change humans to fit technology, fudged our ethics to fit our needs for employment, asked economic life to fit the theories of economists, and asked human to squeeze into some narrative (Taleb 2016, p. 151).

Conservatism (or *situated liberalism*) understands the agent's need for external structures and constraints in order to behave well, cooperate and thrive. These external constraints include laws, institutions, customs, traditions, nations and religions (Marsh and Onof 2008); Doyle and Marsh 2013). This is the broad moral bandwidth, social externalism, we briefly talked of earlier. People who hold this "constrained" view are therefore very "concerned about the health and integrity of these 'outside-the-mind' coordination devices", as Haidt puts it (Haidt 2012, p. 340). So for a conservative, there is always more to morality than harm and fairness (Haidt 2012, p. 149: whereas for the Left fairness is often coextensive with equality while for the Right it means proportionality (Haidt 2012, p. 161).

So what has this to do with reason and rationality? As situated agents "we must be wary of any *individual's* ability to reason ... we should see each individual as being limited, like a neuron. A neuron by itself is not very smart but with a group one gets an emergent system smarter and more adaptive" (Haidt 2012, p. 105).[23] Moreover, good reasoning is an emergent property of a social system" (*ibid*). The worship of abstract reason ("the rationalist delusion") is one of the longest-lived delusions in Western history; as Hayek, Oakeshott, Taleb and others have pointed out.[24] Those who traffic in the delusion are typically what Haidt scathingly calls the "rational caste", a priesthood of philosophers, scientists, politicians, and

---

[23]Bernstein et al. (2012).

[24]Haidt along with Lakoff is the latest in a line of theorists offering a Dual Process Theory (DPT) of cognition, Khaneman (2011) along with Haidt being the two most prominent recent versions. This process comes in a variety of overlapping if not identical binary concerns, notably as propositional knowledge versus tacit knowledge, rationalism versus empiricism, deliberate versus intuitive consciousness, declarative versus procedural, abstract versus hermeneutic, formalized versus traditional (custom, prejudice, convention, habit), fast versus slow thinking and I'd expect more DPT such as these beside. Four points should however be noted. Whoever dual

economists, who usually have a utopian program (Haidt 2012, pp. 33, 94, 95, 103). More disturbingly put: "A fully specified theory of substantive rationality opens the door to despotic requirements, externally imposed" (Nozick 1993, p. 176).

Dispositional Conservatism

Michael Oakeshott once remarked that one can learn more about conservatism from Hume than from Burke (Oakeshott 1991a, p. 435). This somewhat cryptic remark offers a vital clue to fulfilling the promissory note made in earlier writings in which I presented Oakeshott as a situated liberal (Marsh 2010, 2012).

In his essay "On Being Conservative" (OBC) Oakeshott (1991b) famously articulated his self-ascribed conservatism, not as a creed, nor as a doctrine, but as a *disposition*. Despite the term "dispositional conservatism" being one of Oakeshott's most heavily referenced ideas, it has rarely been critically examined.[25] While Oakeshott never quite conceptually unpacks what he means by a disposition,[26] we know that he was profoundly influenced by a Rylean non-Cartesian stance (Marsh 2010).

A disposition is typically a tendency, mood, inclination or potentiality. Within the philosophy of mind, several mental ascriptions have dispositional implications. Ryle claimed that to have a disposition was not to be in a current state but rather to be liable to be in a state when some *other* condition was realized. Ryle's strategy was to show that because mental ascriptions were dispositional they did not involve inner states. This controversial claim meant that ascribing a disposition asserts no more than the truth of a conditional, the upshot being that whatever relation holds

---

process model one cares to take, the positing of these two systems should not lead one to any of the following inferences (Evans 2012):

(a) that there is indeed a sharp duality and that ne'er the twain shall meet; (b) that these two "systems" have definitive brain structure instantiations; (c) that "fast" thinking is intrinsic; (d) that "slow" thinking is intrinsically rational.

The methodological "moral" I take from Kahneman and Tversky's work from the early 70s is not that one should be alert to the supposedly infallible deliverances of intuition but that neither the lay individual nor indeed the expert in any knowledge community, are immune from systematic error. Surely this epistemic insight should be assimilated by social theorists.

[25]A recent exception is Alexander (2016) who attempts a refutation of what he calls Oakeshott's "minimal" and overly "abstract" definition of conservatism. The upshot, on Alexander's account, is that proponents make "conservatism sound like the most natural thing in the world ... [and] unexceptional" (Alexander 2016, p. 20) or as Rescher puts it, conservatism is a "matter of balance", or fine-tuning with the burden of proof on the proponents of change (Marquez 2015), an Aristotelian sense of proportionality. Alexander's point is precisely my point though I have distinguished historically specific, indexical and contextual conservatism (See Rampton 2016); Sutherland (2005) and Turner (2003) are two trailblazers in making the connection between Oakeshott and non-Cartesian cognitive science.

[26]A philosophical unpacking would be in order (see Mumford 1998).

between dispositions and conditionals holds equally between the non-dispositional and conditional. Surely, the argument goes, if dispositions are actual properties, then they exist independently of their manifestations and may still be present even though not made manifest.

The whole of OBC is animated by the effort to defend conservatism (and crucially the conservative disposition with *respect to politics*) without recourse to large-scale religious or metaphysical beliefs and without endorsing a more general traditionalism or communitarianism (Alexander 2015, p. 19). As Capaldi puts it in regard to Hume, a "secular conservatism" that is not in the business of uncritical veneration, but of deploying a sceptical outlook, helps bring to the fore the presuppositions of some ideational domain (Capaldi 1989, p. 311). To get a handle on what might have motivated Oakeshott's remark, we turn to Hume. Oakeshott's position seems to turn on Hume's distinction between "false philosophy" and "true philosophy", the idea being that the former is a corrupt philosophical consciousness that has ironed out the fabric of inherited culture and custom, and is perpetrated by vulgar *rationalistic* philosophical enthusiasms. The latter is "nothing but an attempt to render the inherited and conflicting customs and prejudices of common life as coherent as possible" (Livingston 1995, pp. 154–155; 157; 161). Livingston is not casting "false philosophy" aspersions Burke's way: he's merely saying that Hume had a sharper philosophical eye than Burke, in that he better "distinguished between the legitimate demand for reform and the world inversions of the corrupt philosophical consciousness that informed the French Revolution" (Livingston 1995, p. 159). The true versus false style of philosophizing can be restated, on Livingston's interpretation, as being deeply imbued with a Pyrrhonian doubt and consequent modesty (or as Nyíri 2016 puts it *humbleness*), as opposed to the arrogance, vanity, cynicism and contemptuousness of day-to-day affairs of common life. Livingston's use of the term "delirium" suggests to me that false philosophy is very much a pathological condition, a dogmatic condition that infects zealots of all stripes (Livingstone 1998).

Apriorism in socio-political thinking, or, as Oakeshott's famously put it, "rationalism in politics", is, in its abstraction, too crude and therefore an inappropriate epistemological tool with which to *predictably effect positive* benefits on society. In politics, one is always dealing with a reflective tradition, not an inert pattern of habitual behavior: to reflect a tradition is to "pursue its intimations",[27] as Oakeshott notably and controversially stated. John Coats' view is that, whatever else divides Hume and Oakeshott, they share a similar critique of rationalism, at least within politics and ethics. As Botwinick (2011, p. 148) puts it, 'Oakeshott's concept of "tradition" … straddles the divide between "is" and "ought," … is thus a post-Humean category.' Oakeshott fully assimilates Hume's general proposition of the primacy and relative autonomy of "unreflective custom", with "conscious ratiocination" being the subservient partner while still playing a critical secondary role (Coats 2000, p. 97). In effect, Hume's philosophical psychology offered,

---

[27]See Brannan and Hamlin (2016a) on the stability of practice/convention.

politically speaking, "a conservatism without benefit of mystery" (Wolin 1954, p. 1001) which is fully compatible with Oakeshott's non-metaphysical stance.

In the first half of OBC (§§ 1 and 2) one can recognize the conservative disposition rather well as pertaining to immediate personal circumstances. Oakeshott is making some general observations that would be familiar to theoretical psychology: observations concerning personality traits, habits and attitudes (or, as Oakeshott prefers, "beliefs"). But in respect of politics (§§ 3 and 4 of OBC), the bridge from an individualized dispositional conservatism (§§ 1 and 2 of OBC) is not readily apparent. Herein lies the problem.

In the first half of OBC, Oakeshott sets out what one might term generic conservatism. That is, an individual or a group can be dispositionally conservative independent of any particular substantive political or moral values.[28] *Whether one has a conservative disposition in terms of character, or personality traits in one's private life has nothing, logically or even psychologically, to do with whether one is a conservative in politics.* One can espouse a conservative politics while leading a storm of a private life. *An individualized psychological conservatism can combine with political conservatism, but this is utterly contingent.* The corollary is that one may be a total creature of habit in one's personal life, but still feel the need for extensive political and social change.

Though Oakeshott regards personally held conservatism as a disposition, and a disposition with a significant degree of stability and fixity, he does not think of it as an immutable or irreversible state of character. It has the same fixity and the same reversibility as Aristotelian *hexis*.

Even if one were to regard any given present state of politics as seriously defective one might be unwilling to engage in large-scale reform *not* because one values the current state of affairs, but because one cannot reliably replace it with something better. This is the Oakeshott-Hayek-Popper line of thought, whereby on the grounds of impenetrable social complexity there cannot be a predictive science of politics to drive a radical reconstitution of society. Again, this belief could be held by those with no substantive conservative agenda: it is merely an exercise in descriptive sociology.

Turning to the second half of OBC—conservatism in respect of politics—no tradition of behavior can leave circumstances as they are. In politics one is committed to change, to the remedy of incoherencies. Even Burke said that a society without the means of change is without the means of self-preservation, a claim that is usually completely overlooked by those of a rationalistic stripe and even by many self-identified conservatives. The argument is usually expressed as follows: spontaneous orders à la Hayek and others are sometimes incompatible with, indeed corrosive of, traditional patterns of living. If, as Roger Scruton—whose conservative credentials are beyond reproach—puts it, "in a true spontaneous order the constraints are already there, in the form of customs, laws, and morals" (Scruton

---

[28]Brannan and Hamlin (2016b) distinguish *adjectival* (a value widely-shared across most ideological profiles), *practical* (empirical Pareto-like distribution values) and *nominal* (a distinctively held value)—types of conservatism.

2006, pp. 219–220; Alexander 2015, p. 16), his concern must surely be dissolved.[29] This is pure Hume. Scruton goes onto say that "Hume, Smith, Burke and Oakeshott —have tended to see no tension between a defence of the free market and a traditionalist vision of social order. For they have put their faith in the spontaneous limits placed on the market by the moral consensus of the community" (*ibid*). As mentioned before, the question that has to be asked is what is a traditional order if it is not a spontaneous order?[30]

While one might not necessarily be out of sympathy with some of the rationalists' concerns, their bold activist apriori methodology is tone deaf to the appropriate timing or tipping point to best effect a specific social policy. *Unless the current state of society exactly as it stands is itself a value, then why not remedy incoherencies?* That is why the Married Women's Property Act was justified, as well as female suffrage: women as a social group were equal in all other relevant respects, but in this regard the tradition was incoherent and legal changes had to be made on conservative grounds to remedy the incoherence.[31] What would typically be claimed in the name of human rights can be redescribed as an Oakeshottian "intimation" that was being ignored. The civil rights movement demanded equality for Blacks, another social group, but equality before the law had always been acknowledged under the constitution (15th Amendment, ratified 1870); it's just that there was not due recognition of this equality in respect of voting and education for Blacks. Another notable and very recent example of the remedying of an incoherence for another social group is that of gay rights[32] (inheritance tax equality and

---

[29]Horgan and Timmons' (2007) "morphological rationalism" is the idea information contained in moral principles is already embodied in the structure of an individual's cognitive system, and this morphologically embodied information plays a causal role in the generation of particular moral judgments.

[30]Oakeshott famously took a swipe at Hayek for having rationalist tendencies in that having no plan was just as rationalistic as the central planners (Marsh 2012).

[31]Known primarily for his analytical Marxism Jerry Cohen later wrote: "The conservative attitude that I seek to describe, and begin to defend, in this paper is a bias in favour of retaining what is of value, even in the face of replacing it by something of greater value. I consider two ways of valuing something other than solely on account of the amount or type of value that resides in it. In one way, a person values something as the particular valuable thing that it is, and not merely for the value that resides in it. In another way, a person values something because of the special relation of the thing to that person. There is a third idea in conservatism that I more briefly consider: namely, the idea that some things must be accepted as given, that not everything can, or should, be shaped to our aims and requirements" (Cohen 2011; see Brennan and Hamlin 2016b for a close-grained discussion of Cohen).

[32]Taken thus, I think it way too simplistic to use hot issue terms such as abortion, gay rights, multiculturalism and so on to establish a nuanced conception of the conservative (Tritt et al. 2016). This said, some gays not beholden to the "regressive" Left bemoan the "domestication" of their lifestyle because of legalized marriage. Their view is that gay marriage can be seen as yet another expression of the rationalist's controlling impulse to shrink the transgressive "sandpit" that has long since provided a vital spark to keeping liberal culture dynamic. (See also note 12). Insofar as abortion is concerned, it is first and foremost a philosophical problem, not a political one, as the IYI vulgarian is wont to promote.

equal next-of-kin status). These supposed human rights were historically specific grievances within a tradition of political behavior that needed to be remedied within that very tradition. A tradition that becomes broadly conscious of its being out kilter of with itself, is a tradition best-placed to redress a specific anomaly. (This is not to deny that incommensurability of values can and do occur within and between traditions). This, then, is the counter-charge to human rights claims, which under a wide rhetoric are always related to the socially and historically specific, and can be dealt with in the appropriate local terms.

For Oakeshott, belief is a catch-all for dispositions or dispositional abilities. Given my know-how, I will of course have certain beliefs, but know-how (a capacity or a tendency to act in certain ways) is not identical (or equivalent to) beliefs. Oakeshott reiterates that a political tradition can never be completely coherent. A complex society composed of complex minds, each with their own permutation of beliefs, approvals and disapprovals, preferences and aversions, hopes, fears, anxieties, and skills—will always contain conflicts and tensions. The best we can do is manage and contain them on *reasonable* but *defeasible* grounds. Politics, on this account, is downstream from culture. Thus, complexity issues aside, even the most benevolent of political rationalists is bound to be frustrated.

One potential problem is that one cannot be sure that Oakeshott can draw a firm and tenable distinction between the essential and the incidental within a tradition, or web of belief (see Brennan and Hamlin 2016a). For Oakeshott, everything is connected with everything else. This point reiterated by Livingston is that "there is no suggestion as to what is to be conserved or what the threat is. It is perhaps for this reason that conservatism is often characterized as a disposition rather than a doctrine" (Livingston 1995, pp. 153; 154).

Epilogue

Since I began with Quine, I will end with Quine. While Quine was considered a "political" conservative (Quine 1989, pp. 68, 69, 206–207; White 1999, pp. 125, 272) I'm in accord with David Miller when, with regard to Hume, he writes that Hume's philosophical premises are a necessary though not a sufficient condition for his political stance (Miller 1981, p. 14). This, too, is applicable to Quine. The Quinean web of belief (1970; Vahid 2004) as I understand it, is comprised mainly of beliefs we find ourselves with, and cannot doubt or discard at will.[33] I believe that I am sitting at a computer; I believe that my wife is downstairs; I believe that David Bowie died on the 10th January 2016, and so on and so forth. Any or all of these beliefs could be false, but although there is a degree of voluntariness about a belief, it isn't psychologically possible to discard, modify or adopt a belief at will.

---

[33]For all intents and purposes, I see no difference between Quine's "web of belief" or holism (Quine 1970; Christensen 1994) and Oakeshott's (1933) idealist commitment to coherence.

(This was one of Spinoza's objections to Cartesian doubt. Such doubt, consciously adopted, can only be bogus, or "as if", doubt).

New experiences generate beliefs, some of which (X) are incompatible with some of the beliefs (Y) in the web. In face of this we have to decide, since contradictory or contrary beliefs cannot both be true, whether to hold on to the new beliefs and abandon some of the old, or vice versa.[34] There is no decision-procedure for retaining X and dropping Y, or rejecting X and retaining Y. (The Quine-Duhem thesis sets out why this is so.) Broadly what we actually do, according to Quine, is retain or abandon beliefs in accordance with a judgment of what best restores coherence to the web. A psychological inclination to take the course that least disturbs the web is quite consistent with a radical upheaval or disarrangement of the kind that happened when relativity and quantum theory were adopted. Deep areas of the web were replaced and rewoven.

Quine is at heart a pragmatist (1951). The web is such as to enable us by and large to get along in the world. Sometimes science deeply impinges on it; most of the time it doesn't. The value and justification of the web is the humdrum fact that it is serviceable. Perhaps this aligns it with Oakeshottian tradition, which also is serviceable and also indispensable. There can and could be different webs and different traditions; but without some web or tradition we can't engage with the world as knowers or agents.

Quine doesn't say much about dispositionality in his account of the web of belief, but there is no reason why some beliefs in the web cannot be given a dispositional analysis. One might have supposed that Quine would have rejected dispositions from his austere ontology, but he doesn't (see e.g. Moline 1972). If Quine can accommodate dispositions, there's no reason why he can't accommodate unconscious beliefs, to which Goldman adverts (1979). One can have a dispositional belief of which one isn't aware. That does, however, complicate the process of web revision, since how does one revise—dispositional—beliefs of which one isn't aware?

If we retain the idea of tradition, then I think we're pretty well bound to endorse dispositionality. To adhere to a tradition is to be disposed to make certain "cognitive appraisals" and to behave in certain ways.

1. I don't think Quine or anyone else can get rid of the modal idea of possibility. The actual world is a possible world because it exists and because it can be described without self-contradiction. If it's actual that p then it's possible that p. But I go along with Quine, as with Hume, on necessity and impossibility. All we can say (at most) is that something has always happened or has never happened.

---

[34]A curious schizophrenic-like phenomenon that has typically infected Western intelligentsia of late is satirized by evolutionary behavioural scientist Gad Saad as Ostrich Parasitic Syndrome; that is, the holding of contradictory beliefs manifest as willful ignorance or self-delusion *despite* overwhelming empirical evidence that x is or is not the case (see note 4). Philosophically, this can be roughly recast as a form of Sartrean "bad faith".

2. "Is it true that what a man can do, he can do?" (Gert and Martin 1973). Yes and no—and both are important. Suppose I were to sketch you; I might do a pretty good job. So I can produce a good sketch. But can I reliably produce good sketches? Not at all. It was beginner's luck or a happy chance. If I can do something on one occasion, it does not follow that I can do it reliably or at all on other occasions.

3. Dispositions: Can we apply dispositionals in the same sense in science as in the kind of everyday psychology with which Oakeshott operates?

$$(x)[Cx \rightarrow (Ax \leftrightarrow Bx)]$$

For every x if x is exposed to conditions C then x has property A just in case x exhibits behaviour B. Take this simple illustration: for all butter (x) if butter is exposed to a heat of 200 °C for three hours (C) then butter is solvent (A) if butter melts (B). Can we analyze psychological dispositions in the same way and with the same degree of determinateness?

For every x (a human agent) if x is exposed to conditions C then x is conservative just in case …

One problem is that the scientific disposition, solubility, is single-track as Ryle would say—x melts. I don't think that a disposition such as being conservative is single-track. It's multi-track and can be exhibited in an indefinite variety of—or at least, many—ways. So how to fill in the space after "just in case"? It' s also not clear how to specify "C" in the human agency case.

To conclude, I am not convinced that there is a cast of mind that has either a significant over-preponderance of either rationalist or conservative personality traits. *If such a overly dominant quality of mind did exist, one would have no qualms in designating that as a neurodevelopmental disorder.*[35] Healthy politics, at best, is inevitably a defective experience. The "conservative" stance that the Scottish Enlightenment has been recommending in no way amounts to an unreflective defence or justification of the status quo but is an epistemic plea for a counterposed scepticism, or a Viereckian "animated moderation" (Weinstein 1997) or a "precautionary principle" (Marquez 2015) in matters of wholesale reform to mitigate unintended consequences.[36] This is echoed by Nicholas Rescher who writes that:

---

[35]Some studies have suggested that there is a genetic aspect to an ideological preference (Hatemi et al. 2014) but as the researchers acknowledge, finding bridging laws from bio-chemistry to sociology, is a tall order. If as Tritt et al. (2013) say, emotional arousal is tempered by a conservative stance, then this is clearly dispositional. The same caution should be adhered to in positing a link between media preferences and political orientation (Xu and Peterson 2017).

[36]Hirsh et al. (2010) suggest that personality binary traits such as Conscientiousness versus Openness, Agreeableness versus Politeness, Orderliness versus Openness have some correlation with "conservative" versus "liberal" political outlooks. As I've said, there is only a highly contingent relation if one understands conservatism to be an *epistemic* virtue. This applies to another study by Xu et al. (2016). In an earlier study, Xu et al. (2013, p. 1510) admit to this contingency.

the difference between liberalism and conservatism is not so much one of political—
let alone economic—ideology. Rather, it reflects a difference in temperament, a difference
in attitude regarding the possibilities of the future, which in turn carries in its wake an
attitudinal difference of expectation regarding the ultimate results of contemplated change
(Rescher 2015, p. 442).

Conservatism fortunes as an epistemic stance, has not been well-served because those of a fundamentalist and/or traditionalist stripe have provided the very resilient caricatural fodder (an epistemic hardening of the arteries) to obscure its legitimate strengths. These strengths, as Kristóf Nyíri puts it, are *epistemic* in nature, are *hardly* political if understood in terms of common-sense realism, i.e., *knowledge-conservatism* (Nyíri 2016, p. 444; O'Hara 2016, p. 437). Non-Cartesian cognitive science has come to appreciate this, as has behavioral economics and some parts of computational intelligence (Doyle and Marsh 2013; Frantz and Leslie 2016; Marsh and Onof 2008; Marsh 2010b). Conservatism as an epistemic disposition, is a disposition of *modesty*, a situated liberalism that is vital to the healthy functioning of a culture mediating the perpetual (and necessary) tension between the ying and the yang of the ideal types that Oakeshott identified as the politics of *faith* and the politics of *scepticism*, a mutually valuable epistemic corrective that the broad centre can and should entertain. At perhaps the deepest level of analysis, for Hume, epistemic modesty is a desired intellectual virtue across all domains but most especially with regards to the moral life (i.e., *there can be no a priori knowledge concerning human action*): "Nothing, at first view, may seem more unbounded than the thought of man … But though our thought seems to possess this unbounded liberty, we shall find, upon a nearer examination, that it is really confined within very narrow limits …" (Enquiry § 2).[37]

References

Abel, C., & Marsh, L. (2014). A Danse Macabre of wants and satisfactions: Hayek, Oakeshott, Liberty and Cognition. In G. L. Nell (Ed.), *Austrian economic perspectives on individualism and society: Moving beyond methodological individualism* (pp. 107–140). New York: Palgrave Macmillan.

Alexander, J. (2015). The major ideologies of liberalism, socialism and conservatism. *Political Studies, 63,* 980–994.

Alexander, J. (2016). A dialectical definition of conservatism. *Philosophy, 91,* 215–232.

Hirsh et al. (2013) make a more plausible connection between religiosity and conservatism in that it jibes with Haidt's six drivers of moral psychology. Malka and Soto (2015) again miss this contingent relationship and do not seem to have a very nuanced conception of Left-Right ideological spectrum as per the discussion of ideological morphology in § 2.

[37]I'm indebted to Martin Beckstein, Nick Capaldi, Nathan Cockram, Francesco Di Iorio, Andrew Irvine, Gus DiZerega, David Hardwick, Douglas Livingstone, Kristóf Nyíri, Efraim Podoksik, Geoff Thomas and Seth Vannatta for specific comments and/or broadly related discussion. The usual disclaimers apply.

Amodio, D. M., et al. (2007). Neurocognitive correlates of liberalism and conservatism. *Nature Neuroscience, 10,* 1246–1247.

Beckstein, M. (2017). The concept of a living tradition. *European Journal of Social Theory, 20*(4), 491–510.

Beckstein, M., & Cheneval, F. (2016). Conservatism: Analytically reconsidered. *The Monist, 99,* 333–335.

Bernstein, A., et al. (2012). Viewpoint: Programming the global brain. *Communications of the ACM, 55*(5), 41–43.

Binmore, K. (2011). *Rational decisions (The Gorman lectures in economics).* Princeton: Princeton University Press.

Blum, L. (2013). Political identity and moral education: A response to Jonathan Haidt's. *The Righteous Mind. Journal of Moral Education, 42*(3), 298–315.

Boettke, P. (2017). Hayek's epistemic liberalism. http://oll.libertyfund.org/pages/lm-hayek.

Botwinick, A. (2011). *Michael Oakeshott's skepticism.* Princeton: Princeton University Press.

Brennan, G., & Hamlin, A. (2016a). Practical conservatism. *The Monist, 99,* 336–351.

Brennan, G., & Hamlin, A. (2016b). Conservative value. *The Monist, 99,* 352–371.

Burton, C. M., et al. (2015). Why do conservatives report being happier than liberals? The contribution of neuroticism. *Journal of Social and Political Psychology, 3*(1), 89–102.

Capaldi, N. (1989). *Hume's place in moral philosophy.* New York: Peter Lang.

Christensen, D. (1994). Conservatism in epistemology. *Noûs, 28*(1), 69–89.

Coats, W. J., Jr. (2000). *Oakeshott and his contemporaries: Montaigne, St. Augustine, Hegel, et al.* Selinsgrove and London: Susquehanna University Press; Associated University Presses.

Cohen, G. (2011). Rescuing conservatism: A defense of existing value. In R. J. Wallace, R. Kumar & S. Freeman (Eds.), *Reasons and recognition: Essays on the philosophy of T.M. Scanlon.* Oxford: Oxford University Press.

Corey, D. (2014). Dogmatomachy: Ideological warfare. *Cosmos + Taxis, 3*(1), 60–71.

Dahl, R. (1957). The concept of power. *Systems Research and Behavioral Science, 2*(3), 201–215.

Doyle, M. J., & Marsh, L. (2013). Stigmergy 3.0: From ants to economies. *Cognitive Systems Research, 21,* 1–6.

Duarte, J. L., et al. (2015). Political diversity will improve social psychological science. *Behavioral and Brain Sciences, 38,* 1–13.

Evans, J. (2012). Intuition and reasoning: A dual-process perspective. Questions and challenges for the new psychology of reasoning. *Thinking & Reasoning, 18*(1), 5–31.

Eyal, N., & Tieffenbach, E. (2016). Incommensurability and trade. *The Monist, 99,* 387–405.

Fodor, J. (2003). *Hume variations.* Oxford: Oxford University Press.

Frantz, R., & Leslie, M. (Eds.). (2016). *Minds, models and milieux: Commemorating the centennial of the birth of Herbert Simon.* Basingstoke: Palgrave.

Freeden, M. (1994). Political concepts and ideological morphology. *Journal of Political Philosophy, 2* (2), 140–164.

Froese, T. (2009). Hume and the enactive approach to mind. *Phenomenology and the Cognitive Sciences, 8*(1), 95–133.

Garrett, D. (2006). Hume's naturalistic theory of representation. *Synthese, 152*(3), 301– 319.

Gert, B., & Martin, J. A. (1973). What a man does he can do? *Analysis, 33*(5), 168–173.

Goldman, A. (1979). Varieties of cognitive appraisal. *Noûs, 13,* 23–38.

Green, L. (1988). *The authority of the state.* Oxford: Clarendon Press.

Haack, Susan. (1998). *Manifesto of a passionate moderate: Unfashionable essays.* Chicago: Chicago University Press.

Haack, S. (2012). The embedded epistemologist: Dispatches from the legal front. *Ratio Juris, 25* (2), 206–235.

Haidt, J. (2012). *The righteous mind: Why good people are divided by politics an religion.* New York: Vintage Books.

Haidt, J. (2013a). Moral psychology and the law: How intuitions drive reasoning, judgment, and the search for evidence. *Alabama Law Review, 64,* 867–880.

Haidt, J. (2013b). Moral psychology for the twenty-first century. *Journal of Moral Education, 42* (3), 281–297. https://doi.org/10.1080/03057240.2013.817327.

Haidt, J. (2016). Why concepts creep to the left. *Psychological Inquiry, 27*(1), 40–45.

Hardwick, D., & Marsh, L. (2012a). Clash of the Titans: When the market and science collide. In R. Koppl, S. Horwitz, & L. Dobuzinskis (Eds.), *Experts and epistemic monopolies (Advances in Austrian economics, Volume 17)*. Bingley: Emerald.

Hardwick, D., & Marsh, L. (2012b). Science, the market and iterative knowledge. *Cosmos + Taxis (SIEO), 5*, 26–44.

Haslam, N. (2016). Concept creep: Psychology's expanding concepts of harm and pathology. *Psychological Inquiry, 27*(1), 1–17.

Hatemi, P. K., et al. (2014). Genetic influences on political ideologies: Twin analyses of 19 measures of political ideologies from five democracies and genome-wide findings from three populations. *Behavioral Genetics, 44*(3), 282–294.

Hibbing, J. R., et al. (2014). Differences in negativity bias underlie variations in political ideology. *Behavioral and Brain Sciences, 37*(3), 297–307.

Hirsh, J. B., et al. (2010). Compassionate liberals and polite conservatives: Associations of agreeableness with political ideology and moral values. *Personality and Social Psychology Bulletin, 36*(5), 655–664.

Hirsh, J. B., et al. (2013). Spiritual liberals and religious conservatives. *Social Psychological and Personality Science, 4*(1), 14–20.

Honderich, T. (1991). *Conservatism.* Harmondsworth: Penguin Books.

Horgan, T., & Timmons, M. (2007). Morphological rationalism and the psychology of moral judgment. *Ethical Theory and Moral Practice, 10*(3), 279–295.

Hume, D. (1975). *Enquiries concerning human understanding and concerning the principles of morals,* L. A. Selby-Bigge (Ed.). Oxford: Clarendon Press.

Hume, D. (1978). *A treatise of human nature,* L. A. Selby-Bigge (Ed.). Oxford: Clarendon Press.

Hume, D. (1985). *Essays moral, literary, and political,* E. Miller (Ed.). Indianapolis: Liberty Classics.

Iyer, R., et al. (2012). Understanding libertarian morality: The psychological dispositions of self-identified libertarians. *PLoS ONE, 7*(8), e42366.

Jost, J. T., et al. (2003). Political conservatism as motivated social cognition. *Psychological Bulletin, 129*(3), 339–375.

Kanai, R., et al. (2011). *Political orientations are correlated with brain. Current Biology, 21,* 677–680.

Kaplan, J. T., et al. (2016). Neural correlates of maintaining one's political beliefs in the face of counterevidence. *Scientific Reports.* Article: 39589.

Khaneman, D. (2011). *Thinking, fast and slow.* Toronto: Anchor.

Lakoff, G. (2008). *The political mind: Why you can't understand 21st-Century politics with an 18th-Century brain.* New York: Viking.

Landy, D. (2012). Hume's theory of mental representation. *Hume Studies, 38*(1), 23–54.

Lilienfeld, S. (2017). Microaggressions: Strong claims, inadequate evidence. *Perspectives on Psychological Science, 12*(1), 138–169.

Livingston, D. W. (1995). On Hume's conservatism. *Hume Studies, 21,* (2), 151–164.

Livingston, D. W. (1998). *Philosophical melancholy and delirium: Hume's pathology of philosophy.* Chicago: Chicago University Press.

Malka, A., & Soto, C. J. (2015). Rigidity of the economic right? Menu-independent and menu-dependent influences of psychological dispositions on political attitudes. *Current Directions in Psychological Science, 24*(2), 137–142.

Marquez, X. (2015). An epistemic argument for conservatism. *Res Publica, 22*(4), 405–422.

Marsh, L. (2005). Constructivism and relativism in Oakeshott. In T. Fuller & C. Abel (Eds.), *The intellectual legacy of Michael Oakeshott.* Exeter: Imprint Academic.

Marsh, L. (2010a). Ryle and Oakeshott on the know-how/know-that distinction. In C. Abel (Ed.), *The meanings of Michael Oakeshott's conservatism.* Exeter: Imprint Academic.

Marsh, L. (2010b). Hayek: Cognitive scientist *Avant la Lettre*. In W. N. Butos (Ed.), *The social science of Hayek's 'The Sensory Order'* (*Advances in Austrian Economics*, Volume 13) (pp. 115–155). Bingley: Emerald.

Marsh, L. (2012). Oakeshott and Hayek: Situating the mind. In P. Franco and L. Marsh (Eds.), *A companion to Michael Oakeshott*. University Park: Penn State University Press.

Marsh, L. (2013). Mindscapes and landscapes: Hayek and Simon on cognitive extension. In R. Frantz & R. Leeson (Eds.), *Hayek and behavioural economics*. Basingstoke: Palgrave.

Marsh, L., & Onof, C. (2008). Stigmergic epistemology, stigmergic cognition. *Cognitive Systems Research, 9*(1–2), 136–149.

McArthur, N. (2007). *David Hume's political theory: Law, commerce, and the constitution of government*. Toronto: University of Toronto Press.

Miller, D. (1981). *Philosophy and ideology in Hume's political thought*. Oxford: Clarendon Press.

Moline, J. (1972). Quine on dispositions and subvisible structures. *Mind, 81,* 131–137.

Mumford, S. (1998). *Dispositions*. Oxford: Oxford University Press.

Murray, D. (2017). *The strange death of Europe: Immigration, Identity, Islam*. London: Bloomsbury Continuum.

Nawaz, M. (2016). *Radical: My journey out of Islamist extremism*. Guilford: Lyons Press.

Nozick, R. (1993). *The nature of rationality*. Princeton: Princeton University Press.

Nyíri, K. (2016). Conservatism and common-sense realism. *The Monist, 99,* 441–456.

Oakeshott, M. (1933). *Experience and its modes*. Cambridge: Cambridge University Press.

Oakeshott, (1991a). *Rationalism in politics and other essays*. Indianapolis: Liberty Press.

Oakeshott, (1991b). *On human conduct*. Oxford: Clarendon.

O'Hara, K. (2016). Conservatism, epistemology, and value. *The Monist, 99,* 423–440.

O'Sullivan, N. (1976). *Conservatism*. London: J.M. Dent.

Owen, D., & Davidson, J. (2009). Hubris syndrome: An acquired personality disorder? A study of US Presidents and UK Prime Ministers over the last 100 years. *Brain, 132*(5), 1396–1406.

Percy, W. (1971). *Love in the Ruins*. New York: Farrar Straus and Giroux.

Percy, W. (1991). *Signposts in a Strange Land: Essays*, Patrick Samway (Ed.). New York: Farrar, Straus and Giroux.

Perry, J. (2016). Jesus and Hume among the neuroscientists: Haidt, Greene and the unwitting return of moral sense theory. *Journal of the Society of Christian Ethics, 36*(1), 69–85.

Pinker, S. (2018). *Enlightenment now: the case for reason, science, humanism, and progress*. New York: Viking.

Podoksik, E. (2008). Overcoming the conservative disposition: Oakeshott vs. Tönnies. *Political Studies, 36,* 837–880.

Quine, W. V. O. (1951). Two Dogmas of empiricism. In *From a logical point of view*. New York: Harper & Row.

Quine, W. V. O. (1970). *The web of belief*. New York: McGraw-Hill.

Quine, W. V.O. (1989). *Quiddities: An intermittently philosophical dictionary*. Cambridge, MA: Harvard University Press.

Rampton, V. (2016). The impossibility of conservatism? Insights from Russian history. *The Monist, 99,* 372–386.

Rescher, N. (2015). The case for cautious conservatism. *The Independent Review, 19*(3), 435–442.

Robin, C. (2011). *The reactionary mind: Conservatism from Edmund Burke to Sarah Palin*. Oxford: Oxford University Press.

Russell, B. (1938). *Power: A new social analysis*. London: Allen & Unwin.

Schmitt, C. (1927/1976). *The concept of the political*. Tr. G. Schwab. New Brunswick: Rutgers University Press.

Scruton, R. (2006). Hayek and conservatism. In E. Feser (Ed.), *The Cambridge companion to Hayek*. Cambridge: Cambridge University Press.

Singer, P. (2000). *A Darwinian left: Politics, evolution and cooperation*. New Haven: Yale University Press.

Sutherland, K. (2005). Rationalism in politics and cognitive science. In C. Abel & T. Fuller (Eds.), *The intellectual legacy of Michael Oakeshott* (pp. 263–280). Exeter: Imprint.

Taleb, N. N. (2016). *The bed of procrustes: Philosophical and practical aphorisms*. New York: Random House.

Taleb, N. N. (2018). *Skin in the game: Hidden asymmetries in daily life*. New York: Random House.

Talhelm, T., et al. (2015). Liberals think more analytically (More "WEIRD") than conservatives. *Personality and Social Psychology Bulletin, 41*(2), 250–267.

Tritt, S. M., et al. (2013). Preliminary support for a generalized arousal model of political conservatism. *PLoS ONE, 8*(12), e83333. https://doi.org/10.1371/journal.pone.0083333.

Tritt, S. M., et al. (2016). Ideological reactivity: Political conservatism and brain responsivity to emotional and neutral stimuli. *Emotion, 16*(8), 1172–1185.

Turner, S. (2003). Tradition and cognitive science: Oakeshott's undoing of the Kantian mind. *Philosophy of the Social Sciences, 33*(1), 53–76.

Vahid, H. (2004). Varieties of epistemic conservatism. *Synthese, 141*(1), 97–122.

Verhulst, B., et al. (2012). Correlation not causation: The relationship between personality traits and political ideologies. *American Journal of Political Science, 56*(1), 34–51.

Verhulst, B., et al. (2016). Erratum to "Correlation not Causation: The Relationship between Personality Traits and Political Ideologies". *American Journal of Political Science, 56*(1), 34–51.

von Mises, L. (1949). *Human action: A treatise on economics*. https://mises.org/system/tdf/Human%20Action_3.pdf?file=1&type=document.

Weinstein, M. A. (1997). Peter Viereck: Reconciliation and beyond. *Humanitas, X*(2), 22–40.

Weinstein, E. (2016). Four Quadrant Model. https://everipedia.org/wiki/four-quadrant-model-eric-weinstein/.

White, M. (1999). *A philosopher's story*. University Park: Penn State University Press.

Williams, B. (1985). *Ethics and the limits of philosophy*. Cambridge, MA: Harvard University Press.

Wolin, S. (1954). Hume and conservatism. *The American Political Science Review, 48*(4), 999–1016.

Xu, X., & Peterson, J. B. (2017). Differences in media preference mediate the link between personality and political orientation. *Political Psychology, 38*(1), 55–72.

Xu, X., et al. (2013). Does cultural exposure partially explain the association between personality and political orientation? *Personality and Social Psychology Bulletin, 39*(11), 1497–1517.

Xu, X., et al. (2016). From dispositions to goals to ideology: Toward a synthesis of personality and social psychological approaches to political orientation. *Social and Personality Psychology Compass, 10*(5), 267–280.

Zamboni, G., et al. (2009). Individualism, conservatism, and radicalism as criteria for processing political beliefs: A parametric fMRI study. *Social Neuroscience, 4*, 367–383.

Chapter 13
Do the Social Sciences Need the Concept of "Rationality"? Notes on the Obsession with a Concept

Karl-Dieter Opp

Abstract The present paper criticizes the extensive use of the concept of "rationality" in the social sciences. It is first analyzed what the different issues are when the rationality concept is used. Next treatments of rationality and their major problems by Herbert Simon, Jon Elster, Max Weber, and Raymond Boudon are discussed. It is found, among other things, that these (and many other) authors do not provide arguments for or show the usefulness of the concept of "rationality" that they use. It is further shown that the concept has quite different meanings and is often not clearly defined. Based on this analysis, it is concluded that the concept of rationality is largely superfluous, and it often unnecessarily complicates arguments.

Introduction

"Rationality" is probably one of the most often used concepts in the social sciences. The concept is even part of the name of a major theoretical school, namely the "rational choice approach" which is applied in several social science disciplines such as economics, political science and sociology. There are books with "rationality" in their title such as "The Nature of Rationality" (Nozick 1993). A philosophical school is called "Critical Rationalism." Thus, "rationality" is also a topic in philosophy. If so many scholars from different social science disciplines and philosophy deal with a concept, it is not surprising that it is often ambiguous and, if it is clear, is used in different meanings. If so many scholars employ a concept one would like to know what the reasons are for its use and whether these reasons are acceptable.

K.-D. Opp (✉)
University of Leipzig, Leipzig, Germany
e-mail: opp@sozio.uni-leipzig.de

K.-D. Opp
University of Washington, Seattle, USA

© Springer International Publishing AG, part of Springer Nature 2018
G. Bronner and F. Di Iorio (eds.), *The Mystery of Rationality*,
https://doi.org/10.1007/978-3-319-94028-1_13

These are the questions the present article is concerned with. The first section delineates some basic facts about how to discuss concepts, illustrated with the concept of "rationality."[1] This is useful because many scholars do not seem to be aware of these facts. The remainder of the paper provides a detailed critical analysis of the use of "rationality" in the literature.

How to Discuss the Rationality Concept: Some Basics of the Logic of Concept Formation

One of the pioneers of methodological individualism in sociology is George C. Homans. His discussion of the concept of rationality is well suited to illustrate the different ways of how this concept can be dealt with. In his "Social Behavior" (1974, numbers in parentheses refer to this book) he formulates the "rationality proposition" (43). This is equivalent to value expectancy theory in social psychology (Feather 1982, 1990). It hypothesizes that among the behavioral alternatives a person perceives the behavior is chosen for which the actor *values* the expected consequences relatively positively and *expects* them with a relatively high subjective probability. Thus, the behavior is chosen for which the overall subjective expected utility is highest, from the perspective of the actor.[2] Two questions arise: What does "rationality" mean here? Is the introduction of the concept useful?

Nominal Definitions of "Rationality"

Denoting value expectancy theory as "rationality" proposition means that a person acts rationally when this person engages in subjective utility maximization, as value expectancy theory posits. The first meaning of "rationality" or "rational" thus is:

(1) A person is *rational* = df. the person acts according to value expectancy theory.

The abbreviation "=df" is a shortcut for "is, by definition, to equal in meaning" (Hempel 1952: 3). "Rationality" is thus a shorthand term for a behavior of a person that is consistent with a certain theory. In this definition, the concept "rationality" has a clear meaning. Statement (1) is a *nominal definition*, i.e. a convention suggesting how to use a certain term or expression. In such a definition, the term to be defined (the definiendum, in this case "rational") can always be replaced by the

---

[1]The best English treatise of the logic of concept formation is still Hempel (1952). See also Belnap (1993).

[2]In this essay, the concepts of behavior and action are used interchangeably.

expression that defines the term (the definiens, in this case the sentence after "=df."). Thus, the expression "rationality proposition" can always be replaced by the statements value expectancy theory consists of. Such nominal definitions have several characteristics.

A term such as "rationality" has only the meaning ascribed to it in the nominal definition, i.e. prior meanings are eliminated.

In stipulating a nominal definition, any pre-scientific meaning, including denotative as well as connotative meaning components, is removed. Thus, whatever "rational" or "rationality" has meant before the definition was introduced is, by convention, regarded as non-existent. For example, even if "rational" referred to a conscious behavior in everyday language, this pre-existing denotative meaning is no longer in existence. Furthermore, even if "rational" behavior—whatever this means—is regarded as some behavior that is valued positively (a connotative meaning) this is now, by convention, eliminated.

The definition of "rationality" is not true or false.

A terminological convention can only be more or less *useful*. This holds for typologies as well: they are simply linguistic stipulations.[3]

Criteria for the usefulness of a term

(a) *A definition is useful if it saves time and effort.* Introducing a shortcut for referring to value expectancy theory saves time and effort: one does not need to communicate the statements the theory consists of. One could simply refer to the "rationality proposition" and everybody who has read the definition knows what this term means.

(b) *The shorthand term chosen should have few pre-existing denotative and connotative meanings, but should be easy to memorize.* As was said before, a nominal definition eliminates prior meanings of the term that is defined. This is often not realized by readers. The chance that misunderstandings originate is particularly high if the term to be defined has widely accepted pre-scientific meanings. Even if an author introduces explicitly a definition of a term, readers nonetheless associate with the term prior meanings. An example is the discussion about "reductionism" in sociology. Although "reduction" of a theory is explicitly defined, e.g. by Homans (1974: 12–13), as explanation of a theory by another theory, sociologists associate with "reduction" that the "reduced" theory is to be eliminated. Thus, claiming that "sociology" can be "reduced" to psychology means: "sociological hypotheses can be explained by psychological hypotheses." Nonetheless, this claim is often misunderstood in the sense that sociology should be eliminated. But this is not included in the definition.

---

[3]To many readers all these basic facts about concepts seem obvious. However, there are numerous authors who are not aware of these facts and who cannot distinguish a definition or typology from a theory or empirical proposition. For an illustration see Dubreuil and Grégoire (2013).

In general, it is useful to refrain from using terms with clear prior meanings because this leads to misunderstandings. One extreme possibility is to introduce as a definiendum terms without any prior meanings. Examples would be to use letters from the Greek alphabet such as "psi theory." However, those terms are difficult to remember. It would therefore be more useful to introduce terms that are to some extent related to the meaning ascribed to a term. In regard to "rationality proposition" one could have used simply VET (for "*value expectancy theory*"). This is even shorter than "rationality proposition." One could also speak of the "choice proposition" or of "choice theory" because the theory explains which of the perceived behavioral alternatives is chosen. The term "choice theory" is also used in the literature (see Allingham 2002). This holds for the expression "decision theory" as well. These terms are less burdened with pre-existing meanings and thus elicit fewer misunderstandings than "rationality proposition." However, there are different versions of "choice" or "decision" theory. One should thus add an adjective characterizing the kind of choice or decision theory one wishes to refer to. For example, one version of decision or choice theory is a "wide version" admitting all kinds of preferences, including subjective (and not objective) probabilities, and assuming subjective utility maximization. This is actually value expectancy theory.

(c) *A definition is useful if it is theoretically relevant.* This is the case if it is part of an informative and valid theory. For example, let a proposition state: "If the reward for a behavior increases, the frequency of the behavior increases as well." A first definition of "reward" might be a subjectively valued object that a person receives after he or she has performed a behavior. Research indicates that the proposition in which "reward" is defined in this way is correct. Thus, this definition has theoretical import or, equivalently, is theoretically relevant. Now assume that the following definition of "reward" is proposed: reward is, by definition, an object that is presented to a person after a behavior. This definition is less theoretically relevant because the object might not be valuable to the actor whose behavior is to be explained. Presentation of the object might thus often not increase the frequency of the behavior.

(d) Using the rationality concept might have the *consequence that the respective treatise is attributed scientific importance.* For example, using the expression "axiological rationality" sounds much more scientific and important than simply claiming that actors act according to certain values. It might not be the motivation of an author to use a term in order to be regarded as important, but that might nevertheless be the consequence.

Such extra-scientific motivations or consequences of using a concept need not have anything to do with the scientific value of a concept. It might even be the case, that the reasons for using a concept are purely extra-scientific. Scientifically, thus, the concept may be superfluous.

Whether there are such extra-scientific reasons for using "rationality" and, if so, what these reasons are is never spelled out explicitly. In the present essay, we are interested only in scientific reasons for using concepts and will not further discuss extra-scientific reasons.

Meaning Analysis: How Is Rationality Used by a Group of Speakers?

After stating his "rationality proposition," Homans writes that "we should be aware of the different meanings men have given to the word *rationality*" (1974: 47–48, italics in the original). He then mentions two meanings. The first is that a well-informed observer "views behavior as irrational if he knows its reward is harmful to a man" and, Homans adds, the observer thinks the respective behavior should not be performed (48). This would be the case if a person takes drugs. This is a normative definition. Homans discards this definition because he is not interested in his book in how persons should behave. We will not deal with this normative component either at this point. In regard to the non-normative part of the definition, an irrational behavior is thus an objectively harmful behavior. In everyday language one would accordingly say: a heavy smoker is "irrational."

The second meaning Homans mentioned is that the behavior of a person is irrational "if it is not consciously calculated to get him the largest supply of ... values in the long run" (48). Thus, "rational" behavior refers to behavior that is based on "calculation" of the advantages and disadvantages.

To summarize, Homans hypothesizes two meanings of "rational" or "rationality":

(2) A person acts *rational* means that a well-informed observer knows that rewards the person receives by performing an action are not harmful to him or her in the long run.

(3) A person acts *rational* means that the person calculates the advantages and disadvantages of a behavior.

It is of utmost importance to distinguish the two possible ways of dealing with the concept of "rationality." First of all, the term may be introduced as a *nominal definition*. This happened when Homans defined the expression "rationality proposition." Then Homans turned to existing meanings of the concept. He thus provides a *meaning analysis*. Whereas a nominal definition is a convention about how to use a concept, a meaning analysis yields an *empirical proposition* about the usage of a concept by a certain group of people. A dictionary, for example, does not consist of nominal definitions but aims at describing how people use certain terms or expressions. The two meanings Homans mentioned are hypotheses about how social scientists (or perhaps English speaking people in general) use a concept. These hypotheses can be tested by empirical research and are thus falsifiable. For example, we could present the students of a university (or the population of a country) with a questionnaire where one of the questions is whether the respondents know the meaning of the term "rationality." Those who answer this question in the affirmative could then be presented with a list of meanings like the two mentioned by Homans. The respondents could be asked to mark the meanings they think the concept has. One may in this way measure the extent to which a population associates certain meanings with a concept.[4]

Stipulating nominal definitions and analyses of meaning are related. For example, a concept that is used in a certain way in everyday language may be adopted in a scientific discourse and might be suggested as a nominal definition. If a concept has several meanings such as "rationality," social scientists could select one of its meanings and stipulate this as a nominal definition. For example, meaning (3) above can be found in everyday language. Homans could have stipulated that this meaning is used as a nominal definition. He could thus have *defined* a person as "rational" if the person calculates the advantages and disadvantage of a behavior.

Homans seems to have realized these different ways to deal with the concept of rationality. In regard to normative definitions (i.e. that people *should* behave "rational" in some sense) he states: "... the question whether behavior is rational or not by any of the definitions proposed is irrelevant to our purposes. Our business is with explanation" (1974: 49). Thus, a normative definition is not useful for Homans's purpose which is the explanation of social behavior. In addition, he could have argued that "rationality" in the expression "rationality proposition" is not necessary either because it does not help to solve explanatory questions.

Why does Homans deal with a meaning analysis of the rationality concept when he is actually concerned with defining the concept? If his "business" is explanation, the everyday meaning of rationality is irrelevant. The only issue is how the rationality concept could contribute to the "business" of explaining behavior. For this purpose, a useful nominal definition of "rationality" would be in order and not a meaning analysis.

There are discussions of rationality in the literature in which the questions addressed are: "What *is* rationality?" "What is the *nature* of rationality?" The latter question is even the title of a book by Nozick (1993). In Binmore's book (2009) the Preface begins with the question: "What is rationality?" Such questions are so vague that they can't be answered, let alone be meaningfully discussed. First of all, "is" is not clear. In the question "What *is* rationality?" an author might look for a useful *definition* or might wish to analyze the *meaning* of a term. Second, the concept of "rationality" has many meanings and is often defined in an ambiguous way so that the question of what "rationality" "is" cannot be considered a meaningful question. Such questions should not be part of a scientific book or paper (this is a value judgment of this author!). The ambiguity of "what is" or "what is the nature of" questions can be seen in the following list of sentences with "is." In each sentence, "is" means something different (see Stegmüller 1956):

(1) 2 plus 4 *is* 6. (2) The distance between A and B *is* 50 miles. (3) It *is* prohibited to drink alcohol in public. (4) Every human being *is* mortal. (5) In the English language a husband *is* a man who is married to a woman. (6) We now want to introduce the definition that a norm *is* a statement claiming that something ought to be the case.

---

[4]See Yamagishi et al. (2014) who explore by means of a survey meanings of "rationality."

Thus, "is" can refer to some mathematical relationship (1), to some empirical fact (2), to the description of an existing norm (3), to an empirical law (4), to the meaning of a term (5), and to a definition (6).

The word "nature" is ambiguous as well. The "nature" of rationality may refer to the meaning of the term, to what a fruitful definition could look like, or to how a person should or does behave if she or he is rational. Perhaps the question "What *is* the nature of rationality?" is equivalent to "What is rationality?".

Perhaps some of those who ask the question of the "nature" of rationality wish to find its "'essences' or the 'essential natures' of things—the realities which lie behind the appearances" (Popper 1965: 104; see also Popper 1966: 32). For example, in reading the introduction to Nozick's "The Nature of Rationality" one cannot help to suspect that the aim is to discover some reality that is concealed behind "rationality"—either behind the concept or behind the phenomena the concept refers to. In the latter case, a prerequisite to find these "realities" is a precise definition of what the phenomena are "rationality" refers to. Even if such a definition is provided it is not clear what is meant by "hidden realities" and how they can be discovered. "Rationality" is a linguistic entity that refers to some reality aspects. But what exactly these aspects are is far from clear. A first step in answering these questions should always be to make clear what the concept of rationality means and, thus, which reality aspects one wants to analyze.

Empirical Analyses: To What Extent Are People "Rational"?

There is a third issue related to the rationality concept: Homans assumes that the rationality proposition is an empirically valid propositions. The question Homans is concerned with is thus the *empirical issue* whether people act rationally in the sense defined, i.e. whether people actually behave as value expectancy theory posits. Accordingly, Homans does not only want to provide a definition. Just a linguistic convention does not explain anything. Homans claims that *human beings are rational.* This means that they behave in a certain way.

Many authors who define "rationality," provide meaning analyses or use "rationality" without specifying any meaning also make empirical claims. They usually ask whether individuals are or are not rational. For example, Homans is interested in the question whether a theory—the rationality proposition—is true. But "rational" may refer, as has been said before, to various kinds of properties of actors. For example, if a person is defined as "rational" if he or she calculates the advantages and disadvantages of an action, the empirical question might be whether in general people are "rational" in this sense.

Note that answering empirical questions that include the concept of "rationality" requires to have at least a rough understanding of what the concept is about. Due to the numerous existing meanings of "rationality" and the vagueness of the term simply asking whether people are "rational" is not meaningful.

Rationality Concepts in the Literature

In the remainder of this article we first analyze the work of renowned social scientists who use the rationality concept. The question is which of the problems mentioned before these authors want to solve: is a nominal definition, a meaning analysis or an empirical analysis at issue? We further ask what arguments authors provide for using the rationality concept, and to what extent these arguments are convincing.

There are so many well-known authors who extensively use the rationality concept, that a discussion must be selective. We decided to focus on the work of Herbert A. Simon, Jon Elster, Max Weber and Raymond Boudon. We then discuss briefly a few other authors' treatment of "rationality." Our reading of numerous other writings not discussed in this essay suggests that the flaws of dealing with "rationality" are identical with those of the authors who will be discussed subsequently.

"Rationality" in Herbert A. Simon's "Administrative Behavior"

Herbert A. Simon, Nobel laureate in economics in 1978, is famous for, among other things, his idea of "bounded rationality" (1997, numbers in parentheses refer to this book). This section focuses on the fourth edition of his book "Administrative Behavior" where he provides an extensive discussion of rationality and "bounded" rationality as well. There are numerous other publications where Simon discusses "rationality" and "bounded rationality," but we will only focus on the book from 1997. It will thus not be analyzed to what extent this account is compatible with other writings. Even if his views had changed over time, the account is his book is one possible position that is worth to be discussed.

When Simon first uses the rationality concept in his book he mentions the example of a typist whose behavior of hitting keys is established as a reflex between the respective letter and a particular key. "Here the action is, in some sense at least, rational (i.e. goal oriented), yet no element of consciousness or deliberation is involved" (3). In other words:

> (4) A person is *rational* means, among other things, that a person's behavior is goal oriented.

Simon does not introduce a definition in this sentence. Instead, he describes what "rationality" means. Whether this is the meaning in the English speaking world or only among economists is not said. But this is no problem here because he does not provide a description of the full meaning of the concept, as the expression "in some sense at least" indicates. Thus, the claim is that *there are* speakers who use the term in the sense specified in (4).

In most of his discussion of rationality it seems that Simon is also concerned with a meaning analysis of the concept. For example, he discusses the "correctness" of the decision of an administrator. "It is correct if it [the decision] selects appropriate means to reach designated ends. The rational administrator is concerned with the selection of these effective means. For the construction of an administrative theory it is necessary to examine further the notion of rationality and, in particular, to achieve perfect clarity as to what is meant by the 'selection of effective means'" (72). This argument can be summarized as follows:

(5) (a) A person is (or acts) *rational* means that the person makes a correct decision. (b) A decision is *correct* means that a decision maker selects the appropriate means to reach designated ends.

Simon first defines what a *correct* decision means, viz. if the appropriate means for achieving a given end are selected (see 5b). Then he writes that a *rational* decision maker is concerned with the selection of effective means, i.e. with a correct decision. Is this a statement about the existing meaning of "rational"? Does Simon in this statement presuppose that it is clear what "rational" means? Or does Simon provide a nominal definition of "rational"?

Statements 5 could also be interpreted as a *nominal definition*. Thus, one might specify, that a person is to be defined as "rational" if he or she makes a correct decision, and a correct decision is defined as a decision that selects the appropriate means.

Whatever Simon's statements are about, it is not clear why he uses the term "rational." If it is a nominal definition: is it not sufficient to use "correct"? Why replace "correct" with "rational"? Why should one talk about "rational" instead of "correct" behavior? It is just the replacement of one term ("correct") with another term ("rational"). This makes the whole argument unnecessarily complicated. "Rational" is thus completely superfluous. "Rational" does not add anything to Simon's argument.

Next Simon focuses on means-ends relationships: "ends are ... often merely instrumental to more final objectives. We are thus led to the conception of a series, or hierarchy, of ends. Rationality has to do with the construction of means-ends chains of this kind" (73). What exactly has rationality "to do" with means-ends? Means-ends relationships are often not clear or inconsistent. For example, it is often not clear what exactly the possible or appropriate means for which ends are. "Rational decision-making always requires the comparison of alternative means in terms of the respective ends to which they will lead. As will be seen below ... this means that efficiency—the attainment of maximum values with limited means—must be a guiding criterion in administrative decision" (75). Does this mean that a decision is *defined* as rational if decision makers compare alternative means and ends? It seems more plausible that, again, Simon sees the comparison of means and ends as a *meaning component* of rational behavior. The term "requires" suggests this. It is further important that Simon sets a norm: a "rational" decision maker *should* carry out the comparison. Thus:

(6a) A person is *rational* =df or <u>means</u> that the person *compares* to what extent alternative means lead to the attainment of the person's goals.

(6b) A person *should behave rational* =df or <u>means</u> that the person *should compare* to what extent alternative means lead to the attainment of the person's goals (see the previous sentence 6a).

In this case, a nominal definition would be useful because it would replace the relatively long sentence on the right of (6a). But the term "efficient" instead of "rational"—or any other term such as "thoughtful" or "reasonable"—could be used as well, as Simon himself notes. If Simon's statements are a meaning analysis, the question is not answered why this is necessary in the present context. Simon is concerned with a theory of administrative behavior, and for this purpose the existing meaning of "rationality" is irrelevant, only a definition is needed.

Simon sometimes makes empirical claims about the extent to which people are rational. He writes, for example, that "an assertion that human beings are always or generally rational" is not correct (72). The major goal of Simon is the empirical analysis and explanation of administrative decisions. He never provides a detailed argument specifying why the term "rational" is useful, and in particular why an analysis of the meaning of this term is required for a theory of decision making.

Simon is famous for his concept of *bounded rationality* (see also Simon 1955 where the model is exposed in detail). This expression clearly refers to a set of empirical assumptions explaining human behavior. These assumptions contradict a narrow economic model of man which is "transmuted" to the "person of bounded rationality whom we recognize from everyday life" (118). Assumptions are, for example, that normal people do not maximize utility (i.e. reach the objectively best behavioral outcome) but "satisfice," i.e. choose a course of action that is "good enough" (119). Another assumption is that the "perceived world is a drastically simplified model of the buzzing, blooming confusion that constitutes the real world" (119). Selten (1990: 649) summarizes these ideas as follows: Simon "proposed to replace the idea of utility maximization by a more realistic view of economic behavior involving satisficing and the adaptation of aspiration levels to success and failure. He pointed out that in view of the enormous complexity of the decision tasks confronting firms and consumers optimization transcends human cognitive capabilities."

Simon's ideas about bounded rationality are similar to Homans's "rationality proposition." Simon proposes a "wide" version of "rational choice theory" that is increasingly accepted in the social sciences. Note that Selten's summary does not include the term "rational." "Bounded rationality" is thus a shortcut for a set of assumptions, i.e. a nominal definition. Nonetheless, it seems that the concept carries with it pre-scientific meanings or meanings of a theory that Simon thinks is wrong where "rational" is supposed to refer to, among other things, omniscience and an ability to process complex information. If "rationality" is called "bounded" thus suggests that many of the previous meanings are disregarded. Instead of "bounded rationality" other expressions would do the same service, for example *bounded utility model* or *subjective utility model*. Again, we don't need the term "rational" and perhaps misunderstandings would be avoided without this term.

Jon Elster's Three Requirements for "Rationality"

Elster discusses "rationality" in numerous papers and books. We concentrate on his "Explaining Social Behavior" (2007) which is relatively recent and provides an extensive discussion of "rationality." He addresses "rationality" already at the very beginning of his book. After presenting a long list of puzzles that in his opinion should be explained he turns to one possible theory that could solve these puzzles, viz. rational choice theory (RCT). He asks: "Do many people act on calculations that make up many pages of mathematical appendixes in leading journals? I do not think so" (Elster 2007: 5). Elster advances two allegations in this statement. One is that "rational choice theory" claims that people make complicated calculations. This is clearly an incorrect characterization of this theory (see the section below "Does Rational Choice Theory Assume Rational Behavior?"). RCT is a family of theories. Elster refers to a narrow and now largely outdated version that has been widely criticized, as the previous discussion of Simon's "bounded rationality" shows. A basic flaw in Elster's discussion (see especially Chaps. 11 and 12 in his book) is that he does not clearly distinguish these different versions.

The second allegation in the previous quotation is that "rational" *means* "calculation" (see definition (3) above) and that this is the meaning of "rational" in RCT. This is a typical statement in the literature (see also the introductory section). Elster does not provide a definition, but he describes what he thinks is the meaning of the concept. But it is apparently only one of several meanings. Another meaning Elster refers to is that "a rational agent is one who makes decisions that *make his life go better* as judged by objective criteria such as health, longevity or income" (209, italics in the original). Thus:

(7) A behavior is *rational* <u>means, among other things,</u> the behavior has consequences that make life better.

Elster further makes an empirical claim: "Used in this way the idea would not have any explanatory power. As I have emphasized, *consequences* of a decision cannot explain it" (209, italics in the original—numbers in parentheses refer to Elster 2007).[5] It is not clear what a definition or statement about the meaning of a term such as statement (7) has to do with explanation. Statement (7) does not say anything about explanation: it specifies the meaning of a term and nothing else.

When we ignore this empirical claim it is not clear why a meaning analysis of the term "rationality" is relevant for explaining behavior. If Elster claims that the

---

[5]Incidentally, value expectancy theory that was mentioned before, claims that *perceived* behavioral consequences can very well explain behavior.

meaning of a term is relevant for solving explanatory questions, then it would be useful to analyze other meanings of the term "rational."

We now turn to Elster's discussion of rational choice *theory*. He begins by stating: "Rational-choice theorists want to explain behavior by the bare assumption that agents are rational" (191). Due to the vagueness and inconsistent usage of the term "rational" this is only a meaningful statement if it is clarified what "rational" in rational choice theory means. Elster proposes such a specification of meaning: an action "is" "rational" if it meets three optimality requirements (191–213). (1) The action must be optimal, given the beliefs. That is to say, one must choose "the best means to realize one's desire, given one's beliefs" (202). (2) The beliefs must be "shaped by processing the available evidence using procedures that, in the long run and on average, are most likely to yield true beliefs" (202). (3) There must be "optimal investment of resources—such as time and money—in acquiring more information" (205).

The meaning of these conditions is then discussed in detail. We do not go into this discussion. Instead we ask why Elster needs the term "rational" to describe the three requirements. Instead of asserting

> Rational choice theorists explain behavior by assuming that individuals are *rational*. "*Rational*" means ... (then the three requirements are specified).

one may simply state:

> Rational choice theorists explain behavior by assuming ... (then the three requirements are specified).

Thus, the term "rational" in stating the three "requirements" is not needed at all. This can be seen in our description of the assumptions before: it does not mention "rational." In the original formulation, however, Elster uses "rational" in requirement (2) which begins with: "Rational beliefs are those that are shaped by processing the available evidence ..." (see above). The equivalent formulation is: "it is assumed that beliefs are shaped by processing ..." Again, the term "rational" is not needed at all. On the contrary, without this term time and space are saved. The question arises why Elster uses "rational" in characterizing his assumptions. He does not answer this question.

Are the "requirements" a nominal definition of "rational"? It seems more plausible that Elster uses the existing literature on RCT to state how proponents of this theory define "rational." This is thus a meaning analysis: the requirements summarize how a "rational" person is described by proponents of RCT.

How are the three requirements related to the *theory* of rational choice? Assume that the conditions are met for a particular person P.[6] We can then conclude that the

---

[6] It is highly doubtful that these conditions can *ever* be met. For example, it is unknown what exactly the procedures are that "in the long run and on average, are most likely to yield true beliefs." Furthermore, to determine "optimal investment of resources—such as time and money— in acquiring more information" presupposes an omniscient observer, but observers can also make mistakes.

person "is" rational. Can those properties of the person then be used to *explain* his or her behavior? Due to the first sentence above—"Rational-choice theorists want to explain behavior by the bare assumption that agents are rational"—we would conclude that the person will perform the action for which the requirements are fulfilled. Again, this proposition can be expressed without using the word "rational." One could simply state: P performs behavior B, if the three conditions (i.e. requirements) described are met for the respective behavior. Again, omitting (and not using) "rational" would be less time consuming. Thus, it is difficult to see why the term "rational" is introduced at all.

Another interpretation of the three requirements is possible as well. In another writing Elster holds that the theory of rational choice "is first of all normative, and only secondarily explanatory" (Elster 2009: 14). This interpretation is plausible in the present context because the formulation of the conditions uses a normative language with the term "must." If the requirements are met then this would mean: a person has the property of being (normatively) "rational," i.e. the person conforms to a norm. We will not discuss whether this norm is acceptable, i.e. why a person should behave in the way the requirements specify. For the present context the term "rational" is not needed either for such a normative interpretation of "rational." One could also say, for example, that a person meets the norm of being "reasonable," "efficient" etc. if he or she acts according to the requirements.

We will not discuss in detail the three requirements. Only one note is in order: if the requirements are empirical assumptions that are intended to explain a person's behavior, the question arises whether such a theory is empirically correct. For example, even if assumption (3) is not fulfilled and the actor has completely wrong beliefs, he or she will perform the respective behavior if it is assumed that a subjectively held belief and not an "objectively" correct belief determines a behavior. The actor may then not be "rational" (in the sense that the requirements are met), but empirically the actor will perform the respective behavior. This digression indicates that the term "rational" is not only superfluous in Elster's argument. If this is not accepted, "rational" in Elster's sense is definitely not theoretically relevant because the theory that includes the term is not valid. In using "rational" one might say that actually people behave "irrationally." In conclusion, then, Elster's discussion of rationality confirms a major claim of this essay, namely that we do not need the term "rational," and that dropping this term has advantages.

Max Weber's "Zweckrationalität" and "Wertrationalität"—"Instrumental" and "Value Rationality"

Max Weber (1976: 12–13) distinguishes, among other things, "Zweckrationalität" ("instrumental rationality") and "Wertrationalität" ("value rationality" or, in Boudon's terms, "axiological rationality"—see the next section). Weber's hypothesis is that every type of action can be determined by *instrumental rationality* and *value rationality* (1976: 12, all translations in this section are by KDO).[7] The former type of rationality refers to the realization of the actors' goals. These may also be means to achieve other purposes. "Value rationality" is given if the action is determined by the belief in the intrinsic value of an action, based, for example, on ethical or religious principles, regardless of the success of the action. If such a success it at issue "instrumental rationality" is given.

Weber suggests an empirical proposition: action is, first of all, "determined" by ("bestimmt durch") "instrumental rationality" in the sense that non-normative goals (such as the goal to earn more money) are relevant for performing an action. Secondly, an action may be caused by "value rationality" in the sense that normative goals (such as believing that a robbery to earn more money is morally wrong). Weber *labels* a person whose action is due to the first type of goals, as "instrumentally rational," and a person who acts according to moral principles "value-rational." Weber discusses the effect of these motivations in detail and notes, among other things, that there may also be mixed motives. There are thus two different motivations for action. It seems that Weber introduces *nominal definitions* of instrumental and value rationality. In addition, he suggests *empirical propositions* that these types of rationality have an impact on social action.

These definitions are related to the existing meanings of those concepts in sociology. Weber notes at the beginning of chapter I of the part "Basic Sociological Concepts" ("Soziologische Grundbegriffe") that he only wants to formulate "in a more useful and perhaps more correct way what every empirical sociology actually means if it speaks of the same things" (1976: 1). Nonetheless, the way how Weber introduces the two rationality concepts indicates that he wants to define them in a certain way: he first writes down the concept, then a colon and then the expression which specifies the meaning of the rationality concept. Thus, the nominal definition uses existing meanings and restricts the pre-existing meanings.

Are the two definitions of rationality useful? Couldn't one have avoided them without running into problems? To answer this question let us first put Weber's argument a little more formally:

(8) *Instrumental rationality* of an action =df. an action that is determined by non-normative goals.

(9) *Value rationality* of an action =df. an action that is determined by normative goals.

Proposition 1a: Social action may be instrumentally rational and/or value-rational.

The proposition can be formulated also in the following way:

Proposition 1b: Social behavior is determined by non-normative and/or normative goals.

Thus, if the explanation of behavior is at issue, the definitions of rationality are completely superfluous, as proposition 1b shows. Moreover, the definitions

---

[7]"Wie jedes Handeln kann auch das soziale Handeln bestimmt sein 1. *zweckrational*: ...—2. *wertrational*: ...".

complicate the argument unnecessarily. There is simply no need to introduce the two rationality concepts.

This is confirmed when we try to reformulate the arguments on pages 12–13 where Weber discusses in detail the different possible motivations of social action, including "affective" ("affektuell") and "traditional" motivations, without using the rationality concepts. These pages address the different goals of behavior, and we do not need the rationality concept to describe the numerous motivations that govern social action. For example, we might simply distinguish different *orientations* (or, equivalently, motivations) of an actor such as value orientation, instrumental, traditional and affective orientation. This terminology is used, for example, by Parsons (see, for example, Parsons et al. 1951: 4).

Against this claim it could be argued that Weber wanted to codify and clarify existing sociological terminology, as has been said before. But even if a terminology has been used in the past, the question is whether it is useful and whether one should continue to use it. In particular, why is the distinction between "instrumental rationality" and "value rationality" useful? After finishing the discussion of his "classification of the kinds of action orientation" (13) Weber notes that this classification is not "complete" and that these types of action are "pure" types in the sense that the motivations of action in concrete situations are mixed. His last sentence in his discussion, referring to the classification of the types of action, is: "Their usefulness for *us* can only show their success" (13). But in what sense the concepts might be successful is not discussed.

We thus conclude that the rationality concepts can easily be dispensed with, and that this is possible without any disadvantage. On the contrary, arguments with the referents of the rationality concepts that avoid the rationality concept become less complicated.

Raymond Boudon on Rationality

Raymond Boudon's extensive discussions of rationality is based on Max Weber's types of rationality addressed in the previous section. Boudon distinguishes several kinds of rationality (for a summary see Demeulenaere 2014). The meaning of these concepts becomes especially clear when he criticizes what he thinks are the propositions of RCT (Boudon 1998, 2003, 2009). He lists several "postulates" of the theory. One assumes that "any action is caused by reasons in the mind of individuals (rationality)" (2003: 3). *Rationality* thus means:

(10) A behavior is *rational* =df. a behavior is determined by reasons.

It seems that this is a nominal definition.

"Rationality" is also a property of beliefs. Accepting beliefs such as a scientific theory is not due to costs and benefits, Boudon argues. Beliefs are accepted because they are convincing. Nonetheless, "endorsing the theory is a rational act. ... But here the rationality is cognitive, not instrumental: It consists of preferring the theory that allows one to account for given phenomena in the most satisfying possible way (in accordance with certain criteria). The actor endorses a theory because he or she believes that the theory is true" (2003: 8). This is called *cognitive rationality* "to describe explanations of form: I did so because I believed that 'X is true, likely, plausible, etc.'. So, 'cognitive rationality' would describe the situations where actors believe that 'X is, true, likely, plausible, false, etc.' because, to them, these statements are grounded on reasons which they see as valid and hence are likely to be considered as valid by others" (1996: 124, see also his more formal definition in 2009: 189). Boudon often denotes these reasons as "good" reasons. This may mean that "rationality" or "cognitive rationality" is an attribute of beliefs, but also of persons. To summarize:

(11) A belief is *rational* =df. the belief is accepted due to "good" reasons.

(12) A person is *cognitively rational* =df. a person accepts a belief due to "good" reasons.

Another "postulate" of RCT is that "actors are able to distinguish the costs and benefits of alternative lines of action and that they choose the line of action with the most favorable balance (maximization, optimization)" (2003: 4). One example that allegedly shows the weakness of what Boudon refers to as "RCT" is the voting paradox. A "rational" actor should refrain from voting because the "costs of voting are always higher than the benefits" (2003: 6–7). That is to say, the consequences of going to the polls are not beneficial because a single act of voting in a large group does not contribute to achieving the voters' goals of influencing the kind of government that seizes power. Instrumental rationality refers to achieving the actor's goals (e.g. 2003: 8). This is thus identical (or comes close to) Weber's definition of instrumental rationality (see definition (8) above). But, in addition, Boudon suggests an empirical proposition: the voting example indicates that *instrumental rationality* and, thus, the calculation of costs and benefits, does often not guide behavior.

In Boudon's view, the basic flaw of RCT is that it is restricted to instrumental rationality. It does not take into account that actors follow "nonconsequentialist prescriptive beliefs" (2003: 9). *Axiological rationality* refers to "following non-consequentialist prescriptive beliefs" (2009: 185, for a more formal definition see Boudon 2011: 39). Boudon refers here to Weber's "Wertrationalität" (see definition (9) before).

This is not the place to discuss to what extent Boudon's critique of RCT is acceptable and what the strengths and weaknesses of his own theory are (for this see Opp 2013c, 2014). In this essay the issue is whether the various rationality concepts do add anything to Boudon's theory or to his critique of what he regards as RCT. The previous description of Boudon's arguments indicates already that the different rationality concepts are not only superfluous, they make the theory more complicated and, thus, more difficult to understand.

The arguments for this claim are the following. In later writings Boudon calls his theory "Theory of Ordinary Rationality." This term is not needed. Earlier he uses the expression "the cognitivist model" for his theory (Boudon 1996). So in calling the theory a theory of "rationality" is dispensable.

The major propositions of his theory can be simply formulated in the following way without using the concept of rationality:

> *Boudon's theory*: The major factors that explain (normative as well as descriptive) beliefs are the good reasons actors have for believing something. Reasons actors accept are also the major determinants of social action. These reasons may be instrumental or normative.

This is the bare outline of Boudon's theory (see also Opp 2014). To be sure, he illustrates his ideas by numerous examples, and he shows that these ideas are actually consistent with those of major classic writers, in particular with Max Weber, Émile Durkheim and Alexis de Tocqueville. Nonetheless, the propositions are those in the previous summary.

Boudon's critique of RCT can be formulated without mentioning "rationality." The critique is that RCT neglects norms and values and falsely assumes that individuals always maximize or optimize. It is doubtful whether this critique is tenable (see Opp 2014), but, and this is important in this context, it can be stated without any use of rationality concepts. This is also confirmed by the extensive discussion of rationality concepts by Parri (2014: Chap. 8). To be sure, he uses the term "rationality" extensively himself, but always makes clear what it means. One could have written Chap. 8 in Parri's book without using "rationality."

It is not clear why Boudon does not avoid to use the rationality concept so extensively. Of course, Boudon knows how many other authors use the concept in different ways. One would thus have expected, according to his theory, that he would have given some "good reasons" for using the "rationality" concept so often. What these reasons are, however, is not clear.

A Review of Some Other Rationality Concepts

Let us first return to Homans's "rationality proposition." We find several definitions similar to (1) rather frequently in the literature. Here are some examples.

> (13) A behavior is *rational* =df. the behavior yields the best outcome for an actor (e.g. Rapoport 1966: 142).

The "best" outcome may be defined in different ways, e.g. as the subjectively best outcome from the point of view of an actor or as the objectively best outcome from the point of view of an informed observer. When Luce and Raiffa (1957: 50) present utility theory, which is very similar to value expectancy theory, they state that this theory "is often described as a postulate of rational behavior," i.e. as a postulate of utility maximizing behavior. Here, it seems, "utility maximizing" means objective maximizing, from the viewpoint of an omniscient observer. Mackie (1996: 1005–1006) refers to the game-theoretic accounts of coordination problems that assume "strategic rationality": "'Rationality' means choosing what one wants more over what one wants less, whether self-regarding or other-regarding, given beliefs and constraints." Here subjective well-being is maximized or optimized. All this is consistent with definition (13).

The previous definitions refer to a behavior that is performed and yields the highest outcome. In the following definition "rational" behavior is defined only by a *motivation* to get the best outcome:

(14) A behavior is *rational* =df. the behavior "is motivated by a conscious calculation of advantages, a calculation that, in turn, is based on an explicit and internally consistent value system" (Schelling 1960: 4; similarly Rapoport 1966: 29).

Thus, a behavior is "rational" if it is, among other things, motivated by maximizing utility. This implies that the behavior is "rational" even if the best outcome is not attained. Again, the authors do not provide detailed arguments why they use the concept "rational" in this way.

Definitions of "rationality" that refer to maximizing one's utility are relatively frequent. However, it is often not clear whether subjective or objective maximization (from the perspective of the actor or of an observer) is meant. Coleman (1990: 14), for example, uses "the conception of rationality employed in economics … This conception is based on the notion of different actions … having a particular utility for an actor and is accompanied by a principle that the actor chooses the action which will maximize utility." Does "will" mean that the action actually maximizes utility or that the actor only intends that it yields the highest utility? Another example: it is often said that participating in elections is not "rational." This could mean that the net utility of voting is less than that of abstaining. But it is not clear whether objective (from the viewpoint of an observer) or subjective (from the viewpoint of the actor) maximization is meant.

The previous rationality concepts ascribe rationality to actions, beliefs or actors. "Rationality" often also refers to relations, in this case to a preference ordering (or a relation between preferences). This is a definition widely accepted especially in game theory:

(15) A preference ordering of an individual is *rational* =df. individual preferences are connected and transitive.

A set of objects x and y are *connected* if it holds: either x is preferred to y, or y is preferred to x, or there is indifference between x and y. *Transitivity* means that for any objects x, y, and z, it holds: if x is preferred to y and y is preferred to z, then x is preferred to z.[8]

Why is such a preference ordering called "rational"? Is this term really useful as a shortcut? Assume one wants to discuss the logical or mathematical implications of a connected and transitive preference order. Certainly, there is some saving of time to call this ordering "rational," but the cost is that "rational" is associated with its

---

[8]See, e.g., Arrow (1963): 13, 59, 118. See many other textbooks in social choice, public choice or game theory. A concise description can be found, for example, in Riker and Ordeshook (1973): 19–21. One of the most famous set of axioms of "rationality" are in von Neumann and Morgenstern (1944).

pre-existing meanings. In a book or paper on preference orderings one could, for example, simply introduce the term "CT ordering" for connected and transitive preferences.

There are many other meanings of "rationality," which are sometimes discussed only in passing. Gilboa (2010) reviews several definitions of rationality from the literature and then writes: "I have a preference for a different definition of rationality, which is much more subjective":

> (16) A behavior of a person is *rational* =df. "if this person feels comfortable with it, and is not embarrassed by it, even when it is analyzed for him" (Gilboa 2010: 5).

The last sub-clause (beginning with "even when") means that one feels comfortable despite caveats by others. The author thinks that this definition is useful because a behavior that is "rational" according to the previous definition is rather stable (Gilboa 2010: 5). To be sure, Gilboa is one of the few authors who explicitly states a criterion for the usefulness of the definition: its theoretical import. However, whether just feeling comfortable with a behavior is a strong cause for its stability seems doubtful, unless one explores in detail what exactly the factors are that may change one's feeling of being comfortable. This is then done in the rest of the textbook. It remains open, however, why the behavior is called "rational" and not, for example, just "pleasant" or "agreeable."

There are numerous other meanings of the rationality concept in the social science literature. It is beyond the scope and objective of this paper to provide a complete list of these meanings. Our reading of the literature suggests that the previous definitions are those that are most frequently or at least relatively frequently found in the literatrure.

How to Avoid Talking About Rationality: The Example of the Prisoner's Dilemma

The previous arguments against the usefulness of the rationality concept are not as convincing as demonstrating the plausibility of these arguments with an example. We will do this by analyzing the famous prisoner's dilemma.[9] This is a standard game in game theory. In this theory "rationality" seems to be a key concept. If this is so one would expect that an exposition and discussion of the prisoner's dilemma is not possible without using this concept. Is this really correct? In this section a short description of this dilemma is provided first. Then the behavior of the actors is discussed.

In the following text beginning with the subsection "Assumptions" one may use the rationality terminology. This is struck out and replaced with plain English. This

---

[9]The reader who is not familiar with this game my consult any textbook on game theory such as Luce and Raiffa (1957): 94–102, or Tadelis (2013).

comparison of the text with and without "rational" or "rationality" clearly shows that rationality concepts are superfluous and complicate matters unnecessarily.

The situation of the prisoners. Let two male prisoners A and B be in different cells in a prison. The prisoners cannot communicate and can thus not make joint decisions. The attorney knows that they have committed a certain crime but he thinks he has not enough evidence for a sentence at a trial. The attorney informs the prisoners about the following options and consequences: if they both do not confess the attorney will accuse them of a minor crime that he can prove. Each will be sentenced to 1 year in prison. If they both confess they will be charged for the more severe crime they actually have committed, but they will get a lower sentence due to mitigating circumstances. The sentence will be 5 years in prison for each. But if one confesses and the other not, the confessor will be acquitted and the other one will get the highest possible sentence of 10 years.

Table 13.1a depicts this situation. Each cell first lists the sentence for A, then for B after the slash. For example, let A assume that B does not confess (second column of the table). Given this choice of B: if A does not confess either, A will be sentenced to 1 year in prison; if A confesses he will be acquitted. Now let B confess. Then A is sentenced to 10 years if he does not confess, and to 5 years if he

Table 13.1 The situation of the prisoner's dilemma and the behavioral outcomes

a: Only non-normative goals exist		
Behavioral alternatives of prisoner A	Behavioral alternatives of prisoner B	
	Not confess (Cooperate)	Confess (Defect)
Not confess (Cooperate)	1 year in prison/1 year in prison **3/3**	10 years in prison/Free **1/4**
Confess (Defect)	Free/10 years in prison **4/1**	5 years in prison/5 years in prison **2/2**

b: Normative goals are included		
Behavioral alternatives of prisoner A	Behavioral alternatives of prisoner B	
	Not confess (Cooperate)	Confess (Defect)
Not confess (Cooperate)	1 year in prison/1 year in prison **6/6** $(3 + 3/3 + 3)$	10 years in prison/Free **4/−1** $(1 + 3/4 − 5)$
Confess (Defect)	Free/10 years in prison **−1/4** $(4 − 5/1 + 3)$	5 years in prison/5 years in prison **−3/−3** $(2 − 5/2 − 5)$

Assumptions in Table 13.1b: (1) Each prisoner has a strong internalized norm of cooperation. Conformity to the norm yields an additional benefit of 3, whereas violation of the norm has an additional cost (i.e. negative payoff) of 5. *Examples:* A receives an additional payoff of 3 when he does not confess and when B chooses to cooperate as well; the new payoff in row 2 of Table 13.1b is now 6 (3 + 3). Now let A defect which is the situation in row 3 of the tables. If A has a strong cooperation norm his payoff is diminished by 5, as Table 13.1b shows. The same holds for prisoner B confesses. This situation holds for B as well. For example, let A not confess (see second row of Table 13.1a). If B confesses he will get free. The reader might ignore the numbers in the cells at this point.

Assumptions. Note that some of the following assumptions are described by using "rational" or "rationality." Then these terms are crossed out and followed by other terms with the same meanings. The reader should assess to what extent something is lost if the rationality terminology is avoided and replaced by common language.

Let us now turn to the options of the prisoners which are "confess" or "not confess" or, equivalently, "defect" and "cooperate." Intuitively, "defect" means that the prisoners act so that they do not achieve the best outcomes: if there is bilateral "defection" both prisoners will be sentenced to 5 years, whereas mutual "cooperation" yields 1 year in prison. Another assumption is that the prisoners prefer a lower or no sentence to a higher sentence. Furthermore, the prisoners ~~are rational~~ maximize their utility, i.e. they choose the behavior that is best for them, given the behavior of the other prisoner. The prisoners are egoistic, i.e. they are not concerned with the well-being of others. They are further ~~instrumentally rational~~ concerned with realizing non-normative goals. Their preferences are ~~rational in the sense that they are~~ connected and transitive.

The matrix in Table 13.1a contains numbers. They are the "payoffs" to the prisoners, i.e. their overall valuations of the sentences. It is assumed that the numbers express only the rank order of the valuations. Thus, 4 is better than 3 etc., but numbers do not express how much better an option is valued than another option.

The outcome of the game. How will the prisoners behave? Wouldn't it be ~~rational~~ best for the players to cooperate, because this yields the highest utility for both—each gets a payoff of 3 which is higher than the sum of the payoffs in the other cells? Actually, the ~~rational~~ best behavior for each prisoner is to defect. Let B not confess (defect). It would then be ~~rational~~ best for A to confess because he would get a payoff of 4 which is higher than the payoff of 3 that he would receive if he would not confess. Now let B confess. Again, it would be ~~rational~~ best for A to confess: he is sentenced to 5 and not to 10 years and, thus gets a payoff of 2 instead of 1. The same holds for B. Thus, the outcome (i.e. the joint behavior) is confession. The dilemma is: both would be better off when they cooperated—each would get a payoff of 3, but it is individually ~~rational~~ best to defect (i.e. to defect).

An alternative situation and its consequences. Couldn't there exist a situation in which it becomes ~~rational~~ best for each player to cooperate? So far only the existence ~~instrumental rationality~~ of non-normative goals is assumed. However, individuals are often also or exclusively ~~axiologically rational~~ motivated by values or norms. For example, the prisoners might have strong *internalized norms of cooperation*. This means that an act of cooperation fulfills a norm and is beneficial, whereas defecting is costly. Let us assume that cooperating yields a utility of 3 units, whereas defecting incurs a cost of 5 units. The payoffs in matrix 1a thus change, as is shown in Table 13.1b: if one of the prisoners cooperates, a payoff of 3 is added, if he defects, a payoff of 5 is subtracted. These new payoffs are shown in Table 13.1b. The outcome is different: for each player it is now ~~rational~~ best to cooperate, whatever the choice of the other prisoner is. The overall outcome is thus cooperation. Thus, there may be quite different payoffs in the situation described before. Our example indicates that in a prisoner's dilemma it is not always ~~rational~~ best for the prisoners to defect.

Conclusion. The reader should compare the previous text with and without the rationality concepts. It is clear that the text is much easier to read and also clearer without the term "rational." It is easier to read because everyday language is employed. The term "rational" is not needed at all to understand the argument. It is clearer because, due to the meanings in the literature, the term "rational" has to be explained. Even if this is done, there is still the danger that the reader associates pre-existing meanings with it. Again, the text clearly indicates that "rational" is superfluous and, if it is used, complicates the argument.

There are numerous variations of the game. For example, the game may be repeated. Let "cooperation" be not to pollute, whereas "defection" be to pollute. In this example, the game is repeated because the firm will pollute regularly. Furthermore, more than two players may be involved in a game. In addition, other incentives for choosing defection or cooperation may exist. In regard to pollution, official punishment may be relevant. Finally, there may be more than two options. For example, a player may exit, i.e. leave the situation. Whatever the complications or variations are: we don't need any rationality talk. Every description of the situation and its outcome can be expressed in a straightforward way without using the terms "rational" or "rationality."

Does Rational Choice Theory Assume Rational Behavior?

In discussions of RCT a standard question is: Are people really *rational*, as RCT claims? The answer often is: of course, people are *not* rational, as everybody knows. So this strange theory is plainly false. (By the way, if this argument is correct one wonders what the IQ of proponents of RCT might be!) The previous discussion shows that the question "Are people rational?" is meaningless unless it is specified what "rational" means.

Wouldn't one expect that a theory that has "rational" in its name such as RCT includes a precise definition of "rationality" and claims that people behave "rational" according to this definition? However, this expectation is plainly wrong. Different proponents of RCT use different rationality concept. The previous examples of definitions are to a great deal from proponents of RCT.

So does RCT assume rational behavior and, if so, in what sense? If so many adherents of RCT use different rationality concepts the answer is: *Whether RCT assumes rationality, depends on the version of RCT for which this question is to be answered and on what is meant by "rationality."* We could set up a table in which the first column contains the different rationality concepts. The second column would list the version of RCT which would assert whether people are rational according to a certain rationality concept. Perhaps some cells are empty because no version claims rationality in a given sense.

To illustrate, take definition (13) where a behavior is called *rational* if the behavior yields the best outcome for an actor. This would be assumed by a narrow version of RCT that assumes utility maximization from the perspective of an informed observer. According to a wide version (such as value expectancy theory) people are not rational in this sense: they may misperceive the behavioral consequences and the chosen behavior might make people worse off than a behavior not chosen. Or take definition (14) where a behavior is denoted as *rational* if the consequences are consciously calculated (this is only the first part of the definition and frequently found in the literature). Again, a wide version does not assume rationality in this sense.

Thus, the general answer to the question of whether RCT assumes that people are rational is: (1) First specify what you mean by "rational." (2) Choose the version of RCT for which you wish to answer whether RCT assumes "rationality." (3) Then analyze whether "rationality" as specified in step 1 is assumed by the version of RCT you chose.

Let us return to the typical critique of RCT mentioned before claiming that RCT is wrong because its claim that people act rationally is plainly false. As the previous discussion shows, this claim is simply nonsense. The reason is—let me repeat it, that RCT does not in general posit rationality. Different versions of RCT make different assumptions.

A question that is often asked is: why is rational choice theory called "rational" choice theory? I do not know any historical data that show when and why this name originated and why it spread out. We might speculate that there was a need to characterize the major assumptions of the theory in a general way. And this characterization could have been that the theory assumes "reasonable" behavior. "Reasonable" is a component of the meaning of "rational" so that perhaps the two terms were at first used interchangeably and then "rational" prevailed. Anyway, the label "rational" choice theory is unfortunate because it leads to many misunderstandings which lead to a rejection of the theory.

Conclusion: Should We Dump the Rationality Concept?

The previous analyses have shown that the concept of "rationality" is in general superfluous. We have found that the concept may even unnecessarily complicate arguments. There are only few instances where it could, as a shortcut, save some time and effort. But even in these cases, it can be replaced by other terms. This is meaningful because the concept has strong pre-scientific meanings and will therefore elicit all sorts of misunderstandings. In regard to its theoretical relevance, the concept can be dispensed with as well: if one thinks that rational choice theory is a good theory, "rationality" need not be used, as our previous discussion clearly shows.

Another argument for dumping the concept of rationality is the following. There is an extensive literature where the variables of RCT—different motivations for action such as egoistic or altruistic motivations, internalization of norms or values (normative motivations), different kinds of beliefs—are addressed in great detail and where the concept of rationality is never, or, to be more cautious, extremely rarely used. I refer to the literature in social psychology about learning theory (e.g. Mazur 2006) and value expectancy theory (e.g. Fishbein and Ajzen 2010). One can thus deal with all the phenomena the "rationalists" address without ever using the term "rational." This is also illustrated by a book by Tom R. Tyler, a social psychologist, on "Why People Cooperate" (2013)—a central theme of RCT. Actually, this is the application of a wide version of the rational choice approach. Tyler only uses "rationality" when he refers to RCT (see the index), not in his substantive theoretical argument.

The reader may further scan indexes of any social science book he or she thinks is important for the discipline. How often is the concept "rational" used? For example, in the indexes of Durkheim's "Suicide" (1951) and "Division of Labor" (1964) "rational" does not occur. To be sure, indexes are always incomplete, but if the concept of "rationality" is important in a theory, the author will include it in the index. In Merton's "Social Theory and Social Structure" (1957) the term "rational" cannot be found either. In Parson's "The Social System" (1951) we find an entry "rational action" (pp. 549–550) where he writes that "rationality" may refer to some parts of his theory. But apparently in formulating his theory he did not need the term.

It is further in line with our claim of the uselessness of the rationality concept that a new sociological school, viz. Analytical Sociology, that becomes increasingly popular gets along without the concept of rationality, except in its critique of rational choice theory (see, e.g., Hedström 2005; Demeulenaere 2011). This is remarkable because the basic claims of this school are actually identical with those of rational choice theory (Opp 2013a; see also the controversy: Ylikoski 2013; Manzo 2013; Opp 2013b).

There are certainly readers of this article who teach sociology. I ask these readers: can you imagine that you avoid the concept of rationality in your lectures or seminars, without running in any difficulties? I myself teach sociology, and give, among other things, classes about rational choice theory. In all my talks and lectures where I teach or apply rational choice theory, I never use the concept of rationality, and this does not cause any problem at all.

However, there is one situation where the concept cannot be avoided: in announcing seminars or classes about "rational choice theory" or in papers or talks where this theory is applied. Thus, *the label "rational" is so common as an identifier of rational choice theory that it cannot be avoided in identifying this theory.* In this case, avoiding the concept would lead to misunderstandings. But if the approach to be used is identified, in the specific arguments there is no further need to use this concept.

In order to avoid misunderstandings it should be emphasized that this essay does not imply any critique of the substantive work of the authors discussed. The issue is only their use of the concept of rationality. In regard to the theoretical positions of those authors, it seems that using "rationality" is related to a view that individuals make decisions and that these decisions are based on preferences or goals, and that they are achieved due to the resources and the environment the individuals are faced with. Beyond this basic theoretical orientation there are great differences. For example, some authors hold that in making decisions individuals maximize utility from their subjective perspective, others assume utility maximization from the perspective of an omniscient observer. Furthermore, sometimes non-egoistic preferences (such as altruism) are included, sometimes acting according to such preferences is regarded as "irrational" or "non-rational." Again, our critique does not imply anything about the fruitfulness of the underlying theoretical approach.

The present essay will certainly not change the common use of the concept of rationality. And perhaps some readers are not convinced of the previous arguments. Nonetheless, I hope that a reader who decides to continue to use rationality concepts will observe some *methodological rules* that help to avoid misunderstandings:

1. Whenever one reads a text that uses the "rationality" concept: one should think of the critique presented in his essay. In particular, one should examine whether the meaning of the concept is sufficiently clearly specified and whether reasons are provided for using the concept, and whether these reasons are acceptable.
2. If one decides to use the rationality concept one should define it clearly, provide a detailed argument why it is considered necessary, and consider using other terms.
3. If one wishes that others recognize a work as a contribution to "rational choice theory," then "rational" must be used.
4. If you deal with prior meanings of "rationality" discuss in detail why a meaning analysis is useful for the purposes of your project.

It might be argued against our recommendation to avoid the concept of "rationality" that there are so many vague concepts with different meanings in the social sciences which are regularly used so that there is no convincing reason just to drop the term "rationality." We do definitely not agree with this argument. First of all, it can be doubted that there are many other concept with a similar degree of indeterminacy as "rationality." But even if this is denied it is strange to justify a bad habit with other bad habits. A burglar cannot justify his crime by pointing out that others break into houses as well. Thus, the fact that a concept is widely used is no acceptable argument to imitate a bad practice.

References

Allingham, Michael. (2002). *Choice theory. A very short introduction*. Oxford: Oxford University Press.

Arrow, K. J. (1963). *Social choice and individual values*. New Haven: Yale University Press.

Belnap, N. (1993). On rigorous definitions. *Philosophical Studies, 72*(2/3), 115–146.

Binmore, K. (2009). *Rational decisions*. Princeton and Oxford: Princeton University Press.

Boudon, R. (1996). The 'cognitivist model'. A generalized 'rational-choice-model'. *Rationality and Society, 8*(2), 123–150.

Boudon, R. (1998). Limitations of rational choice theory. *American Journal of Sociology, 104*(3), 817–828.

Boudon, R. (2003). Beyond rational choice theory. *Annual Review of Sociology, 29,* 1–21.

Boudon, R. (2009). Rational choice theory. In B. S. Turner (Ed.) *The New Blackwell companion to social theory* (pp. 179–195). Oxford: Blackwell.

Boudon, R. (2011). Ordinary rationality: The core of analytical sociology. In P. Demeulenaere (Ed.) *Analytical sociology and social mechanisms* (pp. 33–49). Cambridge: Cambridge University Press.

Coleman, J. S. (1990). *Foundations of social theory.* Cambridge, Mass., and London: Belknap Press of Harvard University Press.

Demeulenaere, P. (Ed.). (2011). *Analytical sociology and social mechanisms.* Cambridge: Cambridge University Press.

Demeulenaere, P. (2014). Are there many types of rationality? *Papers. Revista de Sociologia, 99* (4), 515–528.

Dubreuil, B., & Grégoire, J.-F. (2013). Are moral norms distinct from social norms? A critical assessment of Jon Elster and Cristina Bicchieri. *Theory and Decision, 75*(1), 137–152.

Durkheim, É. (1951). (first 1897). *Suicide. A study in sociology.* Glencoe, Ill.: The Free Press.

Durkheim, É. (1964). [1893]. *The division of labor in society.* New York: The Free Press.

Elster, J. (2007). *Explaining social behavior. More nuts and bolts for the social sciences.* Cambridge: Cambridge University Press.

Elster, J. (2009). *Reason and rationality.* Princeton: Princeton University Press.

Feather, N. T. (1982). *Expectations and actions: Expectancy-value models in psychology.* Hillsdale, N.J.: Lawrence Erlbaum.

Feather, N. T. (1990). Bridging the gap betweeen values and actions. Recent applications of the expectancy-value model. In E.T Higgins & R.M. Sorrentino (Eds.) *Handbook of motivation and cognition. Foundations of social behavior* (Vol. 2). New York: Guilford Press.

Fishbein, M., & Ajzen, Icek. (2010). *Predicting and changing behavior. The reasoned action approach.* New York and Hove: Psychology Press.

Gilboa, I. (2010). *Rational choice.* Cambridge, MA: MIT.

Hedström, P. (2005). *Dissecting the social. On the principles of analytical sociology.* Cambridge: Cambridge University Press.

Hempel, C. G. (1952). *Fundamentals of concept formation in empirical science.* Chicago: University of Chicago Press.

Homans, G. C. (1974). *Social behavior. Its elementary forms.* New York: Harcourt, Brace & World.

Luce, R. D., & Raiffa, H. (1957). *Games and decisions. Introduction and critical survey.* New York: Wiley.

Mackie, G. (1996). Ending footbinding and infibulation: A convention account. *American Sociological Review, 61*(6), 999–1017.

Manzo, G. (2013). Is rational choice theory still a rational choice of theory? A response to Opp. *Social Science Information, 52*(3), 361–382.

Mazur, J. E. (2006). *Learning and behavior* (6th ed). New York: Prentice Hall.

Merton, R. K. (1957). *Social theory and social structure.* Glencoe, Ill.: Free Press.

Nozick, R. (1993). *The nature of rationality.* Princeton: Princeton University Press.

Opp, K.-D. (2013a). What is analytical sociology? Strengths and weaknesses of a new sociological research program. *Social Science Information, 52*(3), 329–360.

Opp, K.-D. (2013b). Rational choice theory, the logic of explanation, middle-range theories and analytical sociology: A reply to Gianluca Manzo and Petri Ylikoski. *Social Science Information, 52*(3), 394–408.

Opp, Karl-Dieter. (2013c). Norms and rationality. Is moral behavior a form of rational action? *Theory and Decision, 74*(3), 383–409.

Opp, K.-D. (2014). The explanation of everything. A critical assessment of Raymond Boudon's theory explaining descriptive and normative beliefs, attitudes, preferences and behavior. *Papers. Revista de Sociologia, 99*(4), 481–514.

Parri, L. (2014). *Explanation in the social sciences. A theoretical and empirical introduction.* Soveria Mannelli: Rubbettino.

Parsons, T., Shils, E. A., Allport, G. W., Kluckhohn, C., Murray, H. A., Sears, R. R., et al. (1951). Some fundamental categories of the theory of action: A general statement. In T. Parsons & E. A. Shils (Eds.) *Toward a general theory of action* (pp. 3–29). New York: Harper & Row.

Popper, K. R. (1965). Three views concerning human knowledge. In *Conjectures and refutations. The growth of scientific knowledge* (2nd ed., pp. 97–119). London: Routledge and Kegan Paul.

Popper, K. R. (1966). *The open society and its enemies. Volume 1. The spell of Plato.* London: Routledge & Kegan Paul.

Rapoport, A. (1966). *Two-person game theory. The essential ideas.* Ann Arbor: The University of Michigan Press.

Riker, W. H., & Ordeshook, P. C. (1973). *An introduction to positive political theory.* Englewood Cliffs, N.J: Prentice Hall.

Schelling, T. C. (1960). *The strategy of conflict.* Cambridge, Mass: Harvard University Press.

Selten, R. (1990). Bounded rationality. *Journal of Institutional and Theoretical Economics, 146* (4), 649–658.

Simon, H. A. (1955). A behavioral model of rational choice. *Quarterly Journal of Economics, 69,* 99–118.

Simon, H. A. (1997). (4th ed, first 1945). *Administrative behavior. A study of decision-making processes in administrative organizations.* New York: The Free Press.

Stegmüller, W. (1956). Sprache und Logik. *Studium Generale, 9*(2), 57–77.

Tadelis, S. (2013). *Game theory. An introduction.* Princeton: Princeton University Press.

Tyler, T. R. (2013). *Why people cooperate: The role of social motivations.* Princeton: Princeton University Press.

von Neumann, J., & Morgenstern, O. (1944). *Theory of games and economic behavior.* Princeton: Princeton University Press.

Weber, M. (1976). (6th ed.). *Wirtschaft und Gesellschaft. Grundriß der verstehenden Soziologie, Volume 1.* Tübingen: J.C.B. Mohr (Paul Siebeck).

Yamagishi, T., Li, Y., Takagishi, H., Matsumoto, Y., & Kiyonari, T. (2014). In search of homo economicus. *Psychological Science, 25*(9), 1699–1711.

Ylikoski, P. (2013). The (Hopefully) last stand of the covering-law theory: A reply to Opp. *Social Science Information, 52*(3), 383–393.

Chapter 14
Rationality of the Individual and Rationality of the System: A Critical Examination of the Economic Calculation Problem Over Socialism

Ennio E. Piano and Peter J. Boettke

Abstract We restate Mises' argument about the impossibility of socialist calculation through the lenses of modern developments in microeconomic theory. In so doing, we provide an alternative interpretation of the debate between Austrians and Market Socialists, which we believe should inform economists' understanding of the market process.

Introduction

The socialist calculation debate occupies a central role in the development of modern economic theory and policy. The debate forced the three founding schools of marginalist economics—Walrasian, Marshallian, and Austrian—to realize the full extent of their different methodological and theoretical perspectives (Lavoie 1985; Kirzner 1988; Boettke 2002a, b, 2006). Prior to this development, the major figures in these traditions, such as Mises (1933) and Viner (2013) could argue that the neoclassical schools of economics shared a unified approach. The only difference was in the language employed by each school. After the debate, this picture of a unified neoclassical theory of the price system and the market economy could no longer persist. This stimulated the advancement of economic theory within both the Austrian and the Market Socialist camp. Hayek's theory of the use of knowledge in society, the theory of the entrepreneurial market process, the economics of

E. E. Piano · P. J. Boettke (✉)
George Mason University, Fairfax, USA
e-mail: pboettke@gmu.edu

E. E. Piano
e-mail: epiano@masonlive.gmu.edu

© Springer International Publishing AG, part of Springer Nature 2018
G. Bronner and F. Di Iorio (eds.), *The Mystery of Rationality*,
https://doi.org/10.1007/978-3-319-94028-1_14

information, and mechanism design are among the many modern developments in economics the origin of which can be traced back to the debate over the feasibility of socialist calculation.

The history of the history of the socialist calculation debate is itself illustrative of the development of economic science in the 20th century. Until Don Lavoie's revisionist analysis (1985), the standard treatment of the debate identified five phases. The first phase referred to the development of socialist economics in the Marxian tradition. The second phase was started by Mises' original paper (1920), where he identifies some flaws in contemporary economic theories of socialism but fell short of achieving what the author thought he had done: The theoretical impossibility of the efficient allocation of resources in a socialist economy. The third phase is characterized by the response of socialist economists that economists proved Mises wrong by elaborating various models of market socialism using the tools of neoclassical economics.[1] In the fourth phase, the market socialist response forced Mises' followers F.A. Hayek and Lionel Robbins to retreat from the original anti-socialist position (i.e., that socialism cannot achieve efficiency in the allocation of economic resources) to the more pragmatic stance that socialism would be practically inefficient or at least less efficient than the ideal of an unhampered market economy. The fifth and final phase sees the development, on the part of the market socialists, of a system of trial and error for real-world socialist economies that answered the impracticability argument of Hayek and Robbins.[2]

The purpose of this chapter is to articulate an alternative interpretation of the debate. Building on the revisionist work of Lavoie (1985), we restate Mises' original proposition, contextualizing his "impossibility theorem" in his broader corpus of thought.[3] Our restatement illuminates some of the potential reasons behind the long-lasting incorrect (in our opinion) interpretation of the debate. In Section "The Socialist Proposition and Mises' Challenge", we briefly present the socialist proposition and restate Mises' "impossibility theorem" with a focus on the problems of statics and dynamics, calculation, and rational allocation of resources. In Section "Retreats and Restatements", we outline the market socialist response to Mises' challenge and assess the claim that the Austrians retreated from an "impossibility" to an "impracticability" argument. In this section, we also discuss Oskar Lange's claim that advances in automated computation in the aftermath of the second World War had made the Austrian critique of socialism obsolete. We point out that Lange's critique reveals a fundamental misunderstanding of the notion of calculation and is therefore far from a definitive answer to Mises' original challenge. Section "Conclusion" briefly concludes.

---

[1] These models varied radically from each other and the market socialists themselves spent more time criticizing the specifics of the market socialist models provided by other than responding to the challenges of the anti-socialist side [See, for example, Lerner (1937), which provides a thorough criticism of the competitive socialist framework developed by Durbin (1936)].

[2] The standard recollection of the debate is exemplified by Landauer (1947), Bergson (1948), and Lekachman (1959). See also Schumpeter (1954) and Lavoie (1985, 10–20).

[3] See De Soto (2010) for an overview of the debate in the context of the Austrian tradition.

The Socialist Proposition and Mises' Challenge

The Marxian Vision of the Socialist Society

Mises' "Economic Calculation in the Socialist Commonwealth" was directed toward a very specific form of socialism: The "scientific socialism" developed in the writing of Karl Marx, Friedrich Engels and their followers in Russian and the German speaking world. Breaking with the classical political economy tradition, Marx saw the working of the market economy as inherently unstable, unjust, and destined to be supplanted by a superior form of social organization (i.e., socialism).

In Marx's view, the capitalist system is characterized by production of commodities for market exchange rather than production for direct consumption, thus necessarily leading to specialization between the suppliers of capital services (the capitalists) and the suppliers of labor services (the proletariat). This separation of capital and labor is the root cause of the two major feature Marx attributes to the capitalist system: alienation and exploitation. The institution of private property over land and capital separates man from himself and from nature. He is, in other words, alienated and his ability to satisfy his needs and wants radically constrained. According to Marx, the worker has no control over the overall process of production: he or she does not own either the inputs that go into the production process nor its output, nor has he or she any saying on how production is organized in society.

Alienation has deep repercussions on the working of the capitalist system. The production of commodities for market exchange operates in a decentralized fashion. Marx characterizes this process "not *directly* social, [it] is not the 'offspring of association' which distributes labour internally. Individuals are subsumed under social production; social production exists outside them as their fate; but social production is not subsumed under individuals, manageable by them as their common wealth." (Marx 1953, 158) In a capitalist system, when consumer preferences appear to be different from the ones expected by the producers, "great disturbances may and must constantly occur." (Marx 1967, 315). Rather than an equilibrating tendency, that of the market is a continuous process of disequilibration and social waste (Lavoie 1985, 36; 1983). The anarchy of the capitalist system results in social waste for two reasons. First, not being rationally directed, capitalism must rely on trial and error to achieve some form of mutual coordination. Entrepreneurial action inevitably results in some production plans that, ex-post, are not consistent with consumers' preferences and the relative scarcity of resources. The losses experienced by these entrepreneurs represents the waste in the value of resources due to the decentralized nature of the process. Second, and most importantly, the market process is prone, by its very nature, to cyclical fluctuations. Economic crises such as the Great Depression and the Great Recession result in the idleness of resources which could be employed in the production of valuable goods and services.

Private property and the separation of capital and labor also imply a system in which labor services are sold to the capitalist, which then employs them in the production of goods. Because of their control of the means of production, capitalists can exploit the workers by extracting profits out of the services of their labor. Given that all value must, in Marx's view and arguably in the view of the classical political economists, come from labor, the existence of accounting profit is evidence of the exploitative nature of the capitalist system.[4] Eventually, Marx predicted, these shortcomings will, through the law of the falling rate of profit, lead to its fall and the rise of the socialist society. In the never-ending scramble for profits, capitalists are forced to invest more and more resources in capital goods. But by increasing the ratio capital goods to labor services, by the labor theory of value, the capitalist is also reducing the per-unit surplus of labor they can extract from workers, until no surplus is left and the capitalist way of production collapses onto itself (Marx and Engels 2009).

Marx famously refuted to provide any straightforward plan for the future socialist society and referred with contempt to those "utopian socialists" who adventured themselves in such descriptions. Nonetheless, Lavoie (1985) argues that Marx did not leave us clueless about the organizing principles of a socialist commonwealth[5]:

[M]any of the features of socialism can be inferred from the critique of certain inherent characteristics of capitalism. [...] In many respects, where *Das Kapital* offers us a theoretical "photograph" of capitalism, its "negative" informs us about Marx's view of socialism. Thus, contrary to the standard idea that Marx only talked about capitalism, I am arguing that there is implicit throughout Marx's writings a single, coherent, and remarkably consistent view of socialism. (Lavoie 1985, 30)

Marxian socialism is thus to be seen as maintain all that is good about capitalism, such as its ability to produce unprecedented levels of output and generated technological innovation, while at the same time getting rid of its shortcomings, such as the private property of the means of production and the anarchy of decentralized production for market exchange. The socialist commonwealth will therefore be characterized by the communal ownership of the means of production, the latter being rationally directed by a central planning authority.[6]

Mises' Challenge

Mises' challenge to collectivist economic planning is arguably among the most misunderstood statements in the history of economic theory. The argument,

---

[4]See Blaug (1997) for an overview of Marx's economic thought.

[5]According to Lavoie (1985), this also follows by Marx's adoption of the dialectic method. By the dialectic logic of the laws of historical progression, socialism will eventually take capitalism's place as the organizational framework of human societies. Since socialism is the ultimate and final stage of this development, it must serve as a normative benchmark for all previous stages.

[6]This interpretation is strengthened by the historical evidence that Marx's followers in Russia, among which Bukharih and Preobrazhensky (1979) and Lenin (1920), and Austria (Neurath 1973), attempted to establish socialist regimes in their respective countries (Boettke 1990).

however, is rather straightforward.[7] Without private property of the means of production there can be no exchange and, therefore, no market for the means of production. In the absence of a market for the means of production there can be no money prices for the means of production. Without money prices for the means of production, profit and loss calculation is impossible. Without profit and loss calculation, the system as a whole cannot direct the factors of production towards their most valued uses (Mises 1920, 1949). If socialism is defined as the absence of private property of the means of production, Mises's challenge logically follows: The rational allocation of economic resources under socialism is impossible.

For this "impossibility theorem" to hold, the following conditions must be satisfied:

1. Communal ownership of the means of production under the direction of a planning committee.
2. Consumer goods are freely exchanged at agreed upon terms.[8]
3. The central planner does not have access to any market generated prices of the means of production.[9]

These conditions are important to understand why arguments attempting to disprove Mises on the base of the historical existence of self-proclaimed socialist countries misidentify the target. None of these historical examples fully meet the condition of no exposure to market generated prices of the means of production, as these regimes had access to world markets. Moreover, with the brief exception of War Socialism in Soviet Russia (1918–1921), no socialist country actually attempted the variety of overarching central planning imagined by Marx and his followers (Boettke 1990). Furthermore, Mises never argued that socialism broadly defined was itself impossible.[10] He did not argue that it is impossible for a group of self-identified socialists to take control of the levers of power in one of more countries and use the coercive means of the state to allocate resources in a more or less arbitrary manner. Indeed, Mises witnessed with his own eyes the rise to power of socialist parties in Russia and even his own Austria.

In his article, Mises also makes the following assumptions:

1. The central planner has perfect knowledge about the technological possibilities and tastes of society.
2. The incentives of the central planner are assumed to be perfectly aligned with those of the citizens of the socialist commonwealth. The possibility of opportunistic behavior is ruled out by assumption.

---

[7]See De Soto (2010, 112–119) for an in-depth analysis of Mises' argument.

[8]Here, Mises has in mind Marx's proposal in his *Critique of the Gotha Programme* (Marx 1938). In this scheme, each worker is assigned a number of coupons the value of which is proportional to their contribution to the output of the economy.

[9]One such case would be that of a socialist commonwealth extending over the entirety of mankind.

[10]Caplan (2004) is a modern proponent of this interpretation. See Boettke and Leeson (2005) for a reply to Caplan.

The latter assumption is the cause of a popular criticism of Mises' argument: By focusing on the necessity of prices for rational economic calculation, he down-played the ability of a market economy to align incentives towards the efficient allocation of resources (Buchanan 1969; Demsetz 1982). This criticism is under-standable, and contemporary Austrian economists have done much work to identify the incentive problems of real world socialist regimes (Boettke 1990, 1993, 2000; Anderson and Boettke 1993, 1997)[11] and some [see, for example, White (2012, 39)] have criticized Mises and Hayek for ignoring the issue.

Nonetheless, these criticisms are once again misplaced. First, both Mises and Hayek were fully aware of the incentive problems posed by a centralized control of economic and social activity. In fact, Mises himself refers to the impossibility of aligning the incentives of the managers in a socialist economy: "The management of a socialist concern cannot entirely be placed in the hands of a single individual [the socialist manager], because there must always be the suspicion that he will permit errors inflicting heavy damages on the community. [...] The property owner on the other hand bears responsibility, as he himself must primarily feel the loss arising from unwisely conducted business." (Mises 1920, 121–122) Second, con-textual reading of the debate would probably illuminate why incentives were rel-egated to the background of the calculation debate. Marxian theorists had argued that the rise of socialism will produce a new man, the socialist man, immune from the temptations of opportunistic behavior. Any discussion of motivational issues was therefore destined to be simply dismissed by the socialist side of the debate. Furthermore, at the time, economists themselves saw motivation as the realm of psychology and sociology, not that of economics. For example, this is how Abba Lerner criticizes Durbin (1936) for addressing the possibility of incentive incom-patibilities under capitalism:

> [In the comparison between socialism and capitalism] we must take the theoretical system in both cases – *i.e.*, leaving apart such sociological questions as incentives, etc. [...] He [Mr. Durbin] is, however, guilty of a similar sin [...] when he declares it to be a disad-vantage of capitalist production that the managers of joint-stock companies will reinvest their quasi-rents in their own enterprise, even if the yield is greater elsewhere [...]. This is not an accounting but a personal and sociological problem which may well be even more serious in some forms of socialist economy. (Lerner 1937, 267)

Similarly, Lange writes that "The discussion [of incentives in the socialist economy] belongs to the field of sociology rather than of economic theory and must therefore be dispensed with here." (Lange 1937, 127) Incentive problems were not to be discusses by either side of the debate. To do so would have been a violation of the methodological requirement of value freedom. Thus, Mises cannot be blamed for not stressing this problem enough. He was merely playing by the rules of the game of the scientific conversation of his time.

---

[11]Scholars outside the Austrian tradition have also contributed to the analysis of the consequences of incentive incompatibilities in socialist economies (Bergson 1967; Levy 1990; Shleifer and Vishny 1994).

Statics and Dynamics

Before we move to the next phases of the debate, we want to explore one important point in Mises' original challenge. This is the distinction between the problem of the rational allocation of resources under socialism under static and dynamic conditions. This distinction plays a fundamental role in Mises' argument as well as in its misunderstanding by many market socialists and other neoclassical economists [see, for example, Knight (1936) and Schumpeter (1942)].

According to Mises, economics is the study of human action in the market economy, that is, in a system of social cooperation under the division of labor. Mises defines human action as the purposeful attempt to achieve subjectively identified ends. In equilibrium, or, to use Mises' notion, the final state of rest, there is no human action, and therefore this cannot the object of economic analysis, although the imaginary construction of a static state of affairs can be instrumental in identifying the causal mechanisms that constitute the market process (Mises 1949, 236–237).

Mises' argument about the impossibility of rational economic calculation under socialism must be understood within the broader context of his economics. Under dynamic conditions, central planning cannot rely on the profit and loss mechanism to identify inefficiencies and profit opportunities, and cannot therefore guide production towards the rational allocation of resources. In the market economy, each individual entrepreneur looks at factor prices as determined by the market and at the same time tries to predict at what price he would be able to sell his product or service. On their part, consumers guide production by rewarding those entrepreneurs that have successfully satisfied their wants, and punishing those that have failed to do so. Although calculation in the market is not perfect, the system tends towards the correction of entrepreneurial error and, if it is not perturbed by a change in the underlying economic conditions, will eventually achieve the rational allocation of economic resources in its final state of rest.

Thus, to Mises, the central planner does not face a calculation problem under static conditions, as long as static conditions are defined as an evenly rotating economy reproducing the end results of the market process itself:

> The static state can dispense with economic calculation. For here the same events in economic life are ever recurring; and if we assume that the first disposition of the static socialist economy follows on the basis of the final state of the competitive economy, we might at all events conceive of a socialist production system which is rationally controlled from an economic point of view. (Mises 1920, 109)

Mises is here conceiving that the socialist central planner could guide production under static conditions as long as it commands economic agents to replicate the equilibrium results of the market process.[12] Just a few sentences later, though, he shows that the socialist state would find itself in the paradoxical situation of perpetrating the same production processes, resource allocation, and income distribution of the capitalist system. Any deviation from these would in fact move the economy away from its steady state, but, for Mises' argument, the central planner cannot calculate in disequilibrium. Therefore, the choice of the central planner would be one between the enforcement of the socialist ideal and irrational allocation of resources on the one hand, and inaction and rational allocation of resources on the other.

Retreats and Restatements

The standard account of the socialist calculation debated stated that Mises' challenge was met by neoclassical economists. In the most extreme account, the demonstration by neoclassical economists of the formal similarity of the economic problem under capitalism and socialism was a refutation of Mises. In less extreme accounts, Barone and Pareto's demonstration of the mathematical complexity of the system of equation representing a socialist economy demonstrated Mises was wrong. But in all tellings of the standard account, Lange and Lerner had thoroughly refuted Mises' "impossibility thesis". According to this interpretation, the socialist side demonstrated the theoretical possibility of an efficient allocation of resources in a centrally planned economy, forcing Mises' followers Hayek and Robbins to retreat toward a more pragmatic position:

> Initially the argument focused around the feasibility of one system versus another. It was eventually accepted that both the Lange-Lerner and input-output systems could theoretically answer the economic question about the allocation of resources and manpower. The debated shifted to a dispute over which solution would be the most efficient one. (Goldman 1971, 11)[13]

The Austrians were seen as acknowledging that socialism is possible "in theory" but should still be rejected on the basis of its impracticability and the political danger it poses to liberal-democratic politics. As Lavoie points out, this treatment is biased by two fundamental misunderstandings. First, Austrians and Market Socialists had two different interpretations of the notion of theory. To the latter, to be theoretical meant to analyze equilibrium states (and, in particular, the sufficiency conditions for their existence and the mathematical relationship between their

---

[12]According to De Soto (2010, 106), "Mises himself took very special care to expressly deny that his analysis could be applied to the equilibrium model. This model assumes from the beginning that all necessary information is available and thus, by definition, that the fundamental economic problem socialism poses has been resolved ab initio and in this way, the model leads equilibrium theorists to overlook this problem."

[13]See also Landauer (1947), Bergson (1948), and Sherman (1969).

variables). To the former, on the other hand, all theory must ultimately be directed at understanding the real world. General and partial equilibrium are among the many conceptual devices that can be used to explain the functioning of real world economies. Most importantly, for the Austrians, the analysis of decision making in disequilibrium and the equilibrating process it generates has the same methodological dignity of equilibrium constructs.

Second, the Austrian camp itself committed some important mistakes in conveying their message to the other side. The animosity towards Marxians and Institutionalists in the first decades of the century led them to overestimate the common ground with the Walrasian and Marshallian marginalists. According to Lavoie

> Mises and Hayek tended to take it for granted that ever since the completion of the marginalist revolution of the 1870s, all trained economists had been as subjectivist as they themselves were. (Lavoie 1985, 100–101)

Still in 1933, Hayek's main target was not mathematical economics, but those economists that by refusing "to believe in general laws [of economics were] constitutionally unable to refute even the wildest utopias" (Hayek 1933, 125). One can understand their surprise to see marginalist economics being used to defend (and provide guidance to) a system of central organization of economic activity (Boettke 2006). If they had had a better understanding of the fundamental uniqueness of the Austrian market process approach, they could have expressed less ambiguously their position. The realization of this uniqueness came only as a result of the debate and led Hayek and Robbins to reformulate the original Misesian position by highlighting the themes of the knowledge generating and coordinating properties of the market process in real world economies, the subjective and tacit nature of knowledge, and the role of institutions (Kirzner 1988). Because of the differences in language and emphasis between Mises' original argument on the one hand and Hayek and Robbin's on the other, this restatement gave the impression to those who wanted to hear the message, they interpreted this line of argument as the Austrians' acknowledgement of the victory of the market socialists.[14]

The Market Socialist Response

Mises' challenge produced a radical transformation in the socialist camp. If, before 1920, socialist economics was essentially Marxian, after the publication of Mises'

---

[14]Robbins' position on the matter is illustrative of the Austrian position on the matter. He writes: "There has been much discussion [of the socialist calculation debate] since von Mises first launched his categorical denials, and there are doubtless many who believe that his position has been completely refuted. But that I would venture to question. [...] Far from regarding the von Mises propositions has having been refuted by the march of events. I would hold that the fundamental insight which they contain has been abundantly vindicated." (Robbins 1971, 106–107)

paper socialist embraced the tools and methods of the "bourgeois" neoclassical economics.[15] This change was not restricted to theory and method, but eventually led to the complete abandonment of the very notion of generalized central planning. The socialist response to the impossibility of rational economic calculation under socialism took two forms. The first form is the so-called "equation solving" approach, initially championed by Dickinson (1933) and later resuscitated by Lange (1967). The second form was that of "trial and error" or "competitive solution". Alternative outlines the trial and error or competitive solution to the calculation problem were proposed by Durbin (1936), Lange (1936, 1937), and Lerner (1937, 1938).

Equation Solving

The "equation solving" approach can be summarized as follows. If a socialist economy can be described as a system of equations, then (given some assumptions about the form of these equations) the system has a solution, and in this solution all resources are efficiently allocated. This approach was introduced by two members of the Walrasian school, Pareto and Barone. Unfortunately, the contribution of these authors was misinterpreted by the market socialists and other marginalists as a counterargument to Mises' challenge. Pareto and Barone never argued that a central planner could organize production by solving a system of simultaneous equations (Hayek 1935a). To the opposite, they claimed that this would be an impossible task. In his *Manuel d'Economie Politique*, Pareto argues that a system of equations "has by no means the purpose to arrive at a numerical calculation of prices." Even if the central planner could ever acquire enough information to formulate the equations themselves (something that, the author claims, "is already an absurd hypothesis to make"), the number of equations would be too large for human cognition and "the only means to solve them which is available to human powers is to observe the practical solution given by the market." (Pareto 1927, 233–4) Similarly, Barone defines "fantastic" the idea that production in a socialist regime "would be ordered in a manner substantially different from that of 'anarchist' production." (Barone 1908, 289) Pareto and Barone, but also the Austrian economist Wieser (1914) were merely making an argument about the formal similarity between the required equilibrium conditions under central planning and competition, a point that was never questioned by Mises and his followers (Hayek 1940, 183).

The market socialists understanding of the equation solving argument was radically different. In their idea, if the economy can be represented as a system of equations, then it is logically possible for a central planner to solve it and allocate resources to their most valued uses. Furthermore, some market socialists, including Dickinson (1933), Leontieff (1941), and Lange (1967) argued that central planners could use these equations to guide production in real world economies. Dickinson,

---

[15]The British economist Maurice Dobb was among the few to resist this paradigmatic shift (Dobb 1935).

the earliest proponent of this approach (although he later moved toward the trial and error approach as a superior guide to central planning [Dickinson 1939]), writes that, although "theoretically [...] difficult" the task of the central planner

> could be solved to an approximation sufficiently close for the guidance of the managers of industry, by taking groups of more closely related commodities (composite supply or joint demand) in isolation from other groups. [...] Once the system has got going it will probably be unnecessary to create in this way within the framework of the socialist community a sort of working model of capitalist production. It would be possible to deal with the problems mathematically, on the basis of the full statistical information that would be at the disposal of the S.E.C. [Supreme Economic Council] (Dickinson 1933, 242)

Trial and Error

By the second half of the 1930s, market socialists had moved away from the equation solving approach. The criticisms by Hayek (1935a, b) and Robbins (1935) pushed them to formulate alternative theoretical models of and guidelines for planning in a socialist economy. The main contributors to this literature were Taylor (1929), Durbin (1936), Lange (1936), Lerner (1937, 1938), and Dickinson (1939).[16] These authors saw the adoption of trial and error as the definitive response to the impracticability argument: A process of trial and error can in theory replicate the functioning of the ideal of a perfectly competitive market economy with no need to rely on the assumption of a perfectly omniscient central planner. Not only would such a model of socialism work in theory, the argument goes, but would also outperform real world capitalism by getting rid of such market failures as monopoly power, overproduction, and business cycles. By adopting trial and error, socialism would thus be more equitable and more efficient than the market economy (Lange 1937).

Lange's (1936) is arguably the most sophisticated model of a trial and error approach to central planning. Lange's argument goes as follows. Imagine a socialist economy in which the central planning authority with perfect and complete knowledge of the underlying consumer preferences a well as of the quantity of all economic resources. The central planner would also have "exactly the same knowledge [...] of the productive functions as the capitalist entrepreneurs have." (Lange 1936, 55). Further assume that this economy is characterized by free markets for commodity and labor services, and that the objective of the socialist administration is to satisfy the preferences of consumers. As we discuss above, the formal similarity argument requires that the same equilibrium conditions must emerge in the socialist system as those that would emerge in perfectly competitive markets (Lange 1936, 60–61).

---

[16]MacKenzie (2007) provides a discussion of the trial and error approach and of Mises' preventive criticism of this proposal.

The system achieves these conditions by instructing the mangers of the different industries to minimize average cost and price output such that there is no excess demand. At every period, the central authority announces an index of factor prices to be taken as given by the socialist managers. The latter then mimic the behavior of a neoclassical firm, acquiring units of factors until the marginal cot just equals the output's market price. If, at the end of the period, there is some excess demand (either negative or positive) for some goods and services, the central authority adjusts the index by increasing (decreasing) the price of those factors going into the production of goods in excess demand (supply) (Lange 1936, 63–64, 66–68). Thus, the central authority in a socialist economy serves the same function as the Walrasian auctioneer does in the model of the competitive general equilibrium model. Thanks to what Lange refers to as the "parametric function of prices,"[17] the system is assumed to eventually achieve the optimal allocation of resources.[18]

Lerner (1937) offers a somewhat alternative approach to the issue of efficiency in the socialist economy. Lerner's approach is interesting in many ways. In particular, he claimed that his analysis was derived by the application of the "Austrian" approach, which he thought superior to the Walrasian and Marshallian alternatives in identifying the implications of alternative economic systems (Lerner 1937, 254). According to Lerner, economists on both sides of the debate over the feasibility of socialism misunderstood the fundamental determinant of efficiency in a competitive market and, therefore, in a socialist economy. This confusion led some, including Durbin (1936) to identify two principles that, if closely enforced by the managers of socialist firms, would bring about the efficient allocation of resources.[19] The first principle is that production must be carried on until $P = Min(AC)$, while the second principle requires that $P = MC$.

According to Lerner, this conclusion is deeply mistaken. $P = Min(AC) = MC$ is "merely [an accident] of the state of perfect competition" and not a necessary optimality condition and its application to the management of socialist firms would potentially result in the waste of economic resources (Lerner 1937, 271–272). The only necessary requirement for optimality is that the any unit of output is exactly priced at marginal cost, the latter being defined as "the physical quantity of any factor needed to produce another unit of product multiplied by the price of the factor." (Lerner 1937, 257, 270). Therefore, the socialist economy should not attempt to reproduce the competitive process of market economies. Since the latter's desirability comes entirely from their resulting in marginal cost-pricing, all the

---

[17]See De Soto (2010, 187–188) for an analysis of the distinction between market prices and Lange's interpretation of prices as parameters.

[18]More recent developments in formal economic theory seem to refute Lange's reasoning. See Hurwicz (1973) and Varian (1978).

[19]Lerner (1937, 253) argues that the optimal allocation of resources must be the goal of any rational direction of the economy: "If we so order economic activity of the alternative that is sacrificed, we shall have completely achieved the ideal that the economic calculus of a socialist state sets before itself."

socialist authority has to do is to instruct the managers of factories and industries to set $P = MC$ (Lerner 1937, 271).

The Austrian Restatement

One can only admire the brilliance of Lange's and Lerner's arguments. They consistently and persistently apply the tools developed by economics—the science that studies the mechanism through which a decentralized market economy operates —as organizational principles of a centrally planned economy. The paradoxical result of the use of neoclassical economics for the control of the economy led Mises and Hayek to investigate the originality of the Austrian approach, resulting in the development of a distinctive Austrian theory of the market process [Mises (1922, 1949); Robbins (1935); Hayek (1935a, b, 1940). See also Kirzner (1988) and Boettke (2002a, b)].

The transformation of marginalist economics into a tool for social engineering particularly surprised Hayek, who at the time was teacher and colleague with market socialists such as Durbin, Lange, and Lerner (Boettke 2005, 55, 57). The progression of his thought in the second half of the 1930s is indicative of Hayek's frustration with the tacit and unwarranted assumptions underlying neoclassical economics as it had been used by the market socialists (1935a, b, 1937, 1940, 1945). He also took this intellectual pursuit in two other directions. The first one was an analysis of the political economic implications of the centralized control of economic activities (Hayek 1944), and the potential consequences for the survival of liberal-democratic politics in the West. The second one consisted in a history of the philosophical ideas that influenced the rise of the "scientistic" approach in the social sciences (Hayek 1952).

Boettke (2006) summarizes the result of Hayek's investigation in economic theory and the political economy of socialism in the following way. First, freely operating markets generate and mobilize the tacit and subjective knowledge of millions of individuals.[20] Second, socialism cannot achieve the rational allocation of economic resources because of its inability to mobilize tacit, subjective, and decentralized knowledge. Finally, central planning is incompatible with liberal-democratic principles of governance and, if persistently and consistently exercised, is doomed to result in the rise of an autocratic regime.

The best version of Hayek's reformulation of Mises' argument can be found in two papers: "Economics and Knowledge" (1937) and "The Use of Knowledge in Society" (1945). Although not directly aimed at the market socialists, the content of this paper was clearly influenced by the socialist calculation debate. The former paper focuses mostly on the issue of the assumptions economists make about

---

[20]Modern formulations of the Austrian notion of economic knowledge, see, among others, Lachmann (1986), Boettke (2002, b), and O'Driscoll and Rizzo (2014).

knowledge when modeling market economies. Here, Hayek criticizes the practice of assuming knowledge about preferences, technology, and costs as "given", that is, as "data of the market". To Hayek, this leads to a conceptual confusion between "the objective real facts, as the observing economist is supposed to know them, and [... those] things known to the person whose behavior we try to explain" (Hayek 1937, 39). The importance of this distinction is given by the fact that

> The equilibrium relationship cannot be deduced merely from objective facts, since the analysis of what people will do can start only from what is known to them. Nor can equilibrium analysis start merely from a given set of subjective data, since the subjective data of different people would either compatible or incompatible, that is, they would already determine whether equilibrium did or did not exist. (Hayek 1937, 44)

Thus, while general equilibrium analysis can serve an important role in the analysis of the effect of exogenous shocks to relative prices and the allocation of resources, this cannot, by itself, assess the ability of a system to result in efficient prices and allocations. To accomplish this task, economists must study how the institutional features of the system under study impacts the way in which subjective data are generated and transmitted throughout the economy, and whether these are internally consistent and compatible with the "objective" underlying economic and technological conditions.

Hayek himself provides such an analysis in the latter paper. The agents operating in an economy based on private property and market generated prices are constantly attempting to achieve ends of their own choice. In so doing, they choose the most appropriate means based on each individual's budget constraint and subjective knowledge. These agents do not possess an objective understanding of reality. To interpret their environment, they must rely on subjective, tacit, and often factually inaccurate knowledge. This knowledge can be said to be subjective in two senses. First, it is the result of the interaction of the individual's mind, which is different in its structure from that of any other individual, and physical and social reality. Second, this knowledge can be said to be subjective because it is limited to the specific contingencies of time and space in which the individual operates (Hayek 1945, 80). By this very nature, this knowledge is necessarily dispersed and cannot be collected by one individual or group of individuals without compromising its accuracy.

For a market economy to function, no such concentration of knowledge is required. The price system generated by the competitive market process economizes on the knowledge required by each individual participant to adjust efficiently to each other's behavior, expectations, and to exogenous change. In making their decisions about buying and selling, saving and investing, and so forth, individuals look at market prices, which reflect, although not perfectly, the underlying economic conditions of the economy. An increase in the price of a good leads to the marginal buyers to refrain from consuming it without any need to know the causes of such an increase. Thus, the market process leads to the allocation of resources towards their more valued uses without anyone in the economy knowing what these uses are (Hayek 1945, 85).

Since the knowledge held by the individuals is subjective and limited, there is no a priori reason to believe that is ever going to be correct. The price system (which relies on the feedback mechanism of profits and losses) leads to an adjustment of this knowledge and, as a consequence, of the individual plans that rely upon it.[21] Economic losses signal to the firm that the knowledge and expectations on which its plan was based might be incorrect. For example, the firm might have overestimated the demand of its output from consumers, meaning that it was using an inefficient amount of one or more of the factors of production. The firm can (although not all firms at all time immediately will) adjust to this newly discovered knowledge reduce the level of output or it will eventually go out of business. In either case, the system frees resources from wasteful production processes to be employed in by others in more valuable uses.

Computation and Calculation

Writing more than 30 years after the publication of his model of market socialism, Lange (1967) reflects on the impact of technological innovation in computing for the problem of economic calculation under socialism:

> Were I to rewrite my essay today my task would be much simpler. My answer to Hayek and Robbins would be: so what? Let us put the simultaneous equations on an electronic computer and we shall obtain the solution in less than a second. (Lange 1967, 158)

He then goes on to praise the computer's "unchallenged superiority" to the market process, "a computing device of the pre-electronic age." (Lange 1967, 158–159) The computer is faster, precise, and assured to produce convergence toward a unique equilibrium "by its very construction." (Lange 1967, 160)

Lange recognizes the possibility that the use of linear programming for the planning of real world economies might be impractical in some cases, but, he argues, this should not prevent the central planner to employ computers "for purposes of prognostication" while relying on "the actual working of the market" to fine tune the original algorithm (Lange 1967, 160). Finally, Lange makes the case for the clear superiority of electronic computation over the market in the case of decisions pertaining saving, investment, and development (Lange 1967, 161).

This revival of the "equation solving" argument is telling of the impermeability of the market socialist side to the Austrian challenge in either its original or later forms. Lange's claim that an electronic computer is superior to the market process because the latter will always achieve a stable equilibrium "by construction" is particularly   interesting.   It   betrays   the   persistence   of   a   fundamental

---

[21]In a market economy, prices serve three functions: ex-ante guides to exchange and production; ex-post assessment of previous decisions through the profit and loss mechanism; signals for the existence of discrepancies between ex-ante and ex-post prices, which set in motion the discovery process within the market.

misunderstanding about the nature of the general equilibrium model as a description of the interconnectedness of the economic system and a predictor of the direction of change in response to exogenous shocks. The system of equations is an abstraction that assumes away the economic problem faced by real world economies [the progression towards the mutual consistency of plans of the members of society under conditions of incomplete and asymmetric information and positive transaction costs (Hayek 1948; Demsetz 1982)].

Lange is writing at a time in which the economic profession is moving towards new directions. The new generation of neoclassical economists were starting to critically reject some of the implausible assumptions and results of the market socialist models (including Lange's). Hurwicz (1973) provides a thorough criticism of these models and show how they would not in general meet their stated objectives of an efficient use of decentralized information and the achievement of a stable equilibrium. Bergson (1967) criticizes the market socialists from a different angle. By the 1960s, the neglect of the issue of incentive-compatibility of the socialist economy could not be justified. Socialist planners and managers would face almost insurmountable principal agent problems as well the problem of monopolistic pricing. Furthermore, even assuming these issues away, Bergson claims that no socialist system could process the amount of information necessary to converge toward the equilibrium level of prices. Any change in the underlying economic conditions would require the central planning authority to begin the formulation of the plan from scratch, thus leading to "large and persistent" excess demand (Bergson 1967, 662).

At a more fundamental level, Lange's article betrays as a purely computational interpretation of the market process (Lavoie 1990). If all the market process does is to solve an objectively given system of equations, then the computer (the archetypical "lightning calculator") would of course be superior to the messy process of reciprocal adjustment of the competitive market. Alternatively, the market can be seen (as Austrians do) as the very process through which this knowledge is generated, or "discovered" (Hayek 1948) in the first place. If one adopts this view, the elimination of private property, rivalry, and competition prevents any form of calculation. Even with a perfectly rational[22] individual (or, even better, a computer) in charge, the system could not perform the necessary economic calculation for a rational allocation of economic resources. There is only groping in the dark.

Conclusion

The Calculation debate has been influential in the development of a variety of innovation in economic theory and policy in the 20th century. These includes information economics and mechanism design, but also fields as diverse as linear

---

[22]The word rational is here used in the epistemological rather than praxeological sense.

programming, operations research analysis, and the Austrian theory of the entrepreneurial market process. The debate is also still relevant today. This is reflected in the academic and practical literature. An, while Mises, Hayek, and Robbin's position is better under today than it was between the 1940s and the 1980s, mostly thanks to Lavoie's work, the literature is still filled with misconceptions which can only be addressed via critical reconstruction.

References

Anderson, G. M., & Boettke, P. J. (1993). Perestroika and public choice: The economics of autocratic succession in a rent-seeking society. *Public Choice, 75,* 101–118.

Anderson, G. M., & Boettke, P. J. (1997). Soviet venality: A rent-seeking model of the communist state. *Public Choice, 93*(1–2), 37–53.

Barone, E. (1908). Il ministro della produzione nello stato collettivista. *Giornale degli economisti, 37,* 267–293.

Bergson, A. (1948). Socialism. In *A survey of contemporary economics.* New York: Blakiston.

Bergson, A. (1967). Market socialism revisited. *The Journal of Political Economy, 75,* 655–673.

Blaug, M. (1997). *Economic theory in retrospect.* Cambridge University Press.

Boettke, P. J. (1990). *The political economy of Soviet socialism: The formative years, 1918–1928.* Berlin: Springer Science & Business Media.

Boettke, P. J. (1993). *Why perestroika failed.* UK: Routledge.

Boettke, P. J. (ed) (2000). *Socialism and the market economy: The socialist calculation debate reconsidered* (vol. 9). London: Routledge.

Boettke, P. J. (2002a). *Calculation and coordination: Essays on socialism and transitional political economy.* UK: Routledge.

Boettke, P. J. (2002b). Information and knowledge: Austrian economics in search of its uniqueness. *The Review of Austrian Economics, 15*(4), 263–274.

Boettke, P. J. (2005). On reading Hayek: Choice, consequences and the road to serfdom. *European Journal of Political Economy, 21*(4), 1042–1053.

Boettke, P. J. (2006). Hayek and market socialism. In E. Feser (Ed.), *The Cambridge companion to Hayek.* Cambridge: Cambridge University Press.

Boettke, P. J., & Leeson, P. T. (2005). Still impossible after all these years: Reply to caplan. *Critical Review, 17*(1–2), 155–170.

Buchanan, J. M. (1969). *Cost and choice: An inquiry in economic theory.* Chicago: University of Chicago Press.

Bukharih, N. I., & Preobrazhensky, E. [1919] 1979. *The ABC of communism: A popular exploration of the program of the communist party of Russia* (E. Paul & C. Paul., Trans.). Ann Arbor: University of Michigan Press.

Caplan, B. (2004). Is socialism really "impossible"?. *Critical Review, 16*(1), 33–52.

Demsetz, H. (1982). *Economic, legal, and political dimensions of competition* (Vol. 4). North Holland.

De Soto, J. H. (2010). *Socialism, economic calculation and entrepreneurship.* Edward Elgar.

Dickinson, H. D. (1933). Price formation in a socialist community. *The Economic Journal,* 237–250.

Dickinson, H. D. (1939). *The economics of socialism.* Oxford: Oxford University Press.

Dobb, M. (1935). Economic theory and socialist economy: A reply. *The Review of Economic Studies, 2*(2), 144–151.

Durbin, E. F. (1936). Economic calculus in a planned economy. *The Economic Journal, 46*(184), 676–690.

Goldman, M. I. (1971). *Comparative economic systems: A reader.* Random House.

Hayek, F. A. (1933). The trend of economic thinking. *Economica, 40,* 121–137.

Hayek, F. A. (1935a). The nature and history of the problem. In *Collectivist economic planning* (pp. 1–40). London: Routledge.

Hayek, F. A. (1935b). The present state of the debate. In: *Collectivist economic planning* (pp. 201–243). London: Routledge.

Hayek, F. A. (1937). Economics and knowledge. *Economica, 4*(13), 33–54.

Hayek, F. A. (1944) *The road to serfdom.* London: Routledge.

Hayek, F. A. (1945). The use of knowledge in society. *The American Economic Review, 35*(4), 519–530.

Hayek, F. A. (1948). *Individualism and economic order.* Chicago: University of Chicago Press.

Hayek, F. A. (1952). *The counter-revolution of science: Studies on the abuse of reason.* Glencoe, Illinois: The Free Press.

Hayek, F. V. (1940). Socialist calculation: The competitive solution'. *Economica, 7*(26), 125–149.

Hurwicz, L. (1973). The design of mechanisms for resource allocation. *The American Economic Review, 63*(2), 1–30.

Kirzner, I. M. (1988). The economic calculation debate: lessons for Austrians. *The review of Austrian economics, 2*(1), 1–18.

Knight, F. H. (1936). The quantity of capital and the rate of interest: I. *The Journal of Political Economy, 44,* 433–463.

Lachmann, L. M. (1986). *The market as an economic process.* Wiley-Blackwell.

Landauer, C. (1947). *The theory of national economic planning.* California: University of California Press.

Lange, O. (1936). On the economic theory of socialism: part one. *The review of economic studies, 4*(1), 53–71.

Lange, O. (1937). On the economic theory of socialism: part two. *The Review of Economic Studies, 4*(2), 123–142.

Lange, O. (1967). The computer and the market. In C. H. Feinstein (Ed.), *Socialism, capitalism, and economic growth: Essays presented* (pp. 158–161). Cambridge: Cambridge University Press.

Leontief, W. (1941). *The structure of the american economy 1919–1929.* Oxford University Press.

Lavoie, D. (1983). Some strengths in Marx's Disequilibrium theory of money. *Cambridge Journal of Economics, 7,* 55–68.

Lavoie, D. (1985). *Rivalry and central planning: the socialist calculation debate reconsidered.* Cambridge: Cambridge University Press.

Lavoie, D. (1990). Computation, incentives, and discovery: the cognitive function of markets in market socialism. *The Annals of the American Academy of Political and Social Science, 507,* 72–79.

Lekachman, R. (1959). *A history of economic ideas.* Harper.

Lenin, V. I. (1920). *State and revolution.* Sidney: Australian Socialist Party.

Lerner, A. P. (1937). Statics and dynamics in socialist economics. *The Economic Journal, 47*(186), 253–270.

Lerner, A. P. (1938). Theory and practice in socialist economics. *The Review of Economic Studies, 6*(1), 71–75.

Levy, D. M. (1990). The bias in centrally planned prices. *Public Choice, 67*(3), 213–226.

MacKenzie, D. W. (2007). Trial and error in the socialist calculation debate. *History of Political Economy.*

Marx, K. (1938). *Critique of the Gotha programme.* New York: International Publishers.

Marx, K. (1953). *Grundrisse.* Pelican Books.

Marx, K. (1967). *Capital: A critique of political economy* (Vol. 2). International Publishers.

Marx, K., & Engels, F. (2009). *The economic and philosophic manuscripts of 1844 and the Communist manifesto.* Prometheus Books.

Mises, L. V. (1920). Economic calculation in the socialist commonwealth. In Hayek, F. A. (ed.), *Collectivist economic planning* (pp. 30–31). London: Routledge.

Mises, L. V. [1922] 1981. *Socialism: An economic and socialogical analysis*. Indianapolis: liberty Press.

Mises, L. V. [1933] 1960. *Epistemological problems of economics*. Auburn: Ludwig von Mises Institute.

Mises, L. V. [1949] 2007. *Human action: A treatise on economics*. Indianapolis: Liberty Fund.

Neurath, O. (1973). Through war economy to economy in kind. In *Empricism and Sociology*. Boston.

O'Driscoll Jr, G. P., & Rizzo, M. (2014). *Austrian economics re-examined: The economics of time and ignorance* (Vol. 33). London: Routledge.

Pareto, V. [1927] 1971. *Manual of political economy*. New York: Kelley.

Robbins, L. (1935). *The great depression*. London: Macmillan & Co.

Robbins, L. (1971). *An economist's autobiography*. Macmillian.

Schumpeter, J. A. (1942). *Socialism, capitalism and democracy*. New York: Harper and Brothers.

Schumpeter, J. A. (1954). *History of economic analysis*. London: Routledge. 2006.

Sherman, H. J. (1969). *The soviet economy*. Little, Brown.

Shleifer, A., & Vishny, R. W. (1994). The politics of market socialism. *The journal of economic perspectives, 8*(2), 165–176.

Taylor, F. M. (1929). The guidance of production in a socialist state. *The American Economic Review, 19*, 1–8.

Varian, H. (1978). *Microeconomic analysis*. WW Norton.

Viner, J. (2013). *Jacob Viner: Lectures in economics 301* (Vol. 301). New Jersey: Transaction Publishers.

White, L. H. (2012). *The clash of economic ideas: the great policy debates and experiments of the last hundred years*. Cambridge University Press.

Wieser, F. V. (1914). *Social economics*. London: Allen and Unwin. 1927.

Chapter 15
Rationality and Interpretation in the Study of Social Interaction

Emmanuel Picavet

Abstract The widespread use of the rational choice model (or rational-choice methodology) in various fields outside economic theory, such as politics and sociology, is often understood and described as an import from the methodology of economics. The associated ideas about the "imperialism" of economic methodology cannot be taken for granted; neither should we leave the equation of rationality and end-means association unquestioned. The core of the optimizing model of rationality is consistent decision-making, when the guiding values are assumed to be given or somehow accessible. Interpretative issues can be important in economic matters, even though the role of interpretation is often masked in economics as a discipline because the goals are taken to be obvious. Indeed, it is argued that modeling tasks are at root closely connected with interpretative tasks. The role of interpretation is even more conspicuous outside economics, whenever complex motivational guidelines are in operation. Interpretation can hardly be separated from modeling tasks when human action is understood or explained, starting from good reasons or the availability of appropriate reasoning. This has implications for the understanding of norm-based choices of actions in methodological individualism.

Introduction

Rationality is both a philosophical notion and a modeling assumption or theoretical hypothesis in the social sciences.[1] This duality is quite significant because it secures a connection between rationality as such (with a number of things we know about it) and methodological guidance. This is particularly apparent in the so-called "rational choice model" which has complex ramifications in the humanities and in

---

[1] I would like to thank Alicia Dorothy Mornington-Engel for her very helpful comments.

E. Picavet (✉)
Université Paris 1 Panthéon-Sorbonne, Paris, France
e-mail: emmanuel.picavet@univ-paris1.fr

© Springer International Publishing AG, part of Springer Nature 2018
G. Bronner and F. Di Iorio (eds.), *The Mystery of Rationality*,
https://doi.org/10.1007/978-3-319-94028-1_15

social science. "Rational" in this phrase is not just a nickname for methodological adequacy: it refers to (human) practical rationality. On the other hand, the reference to modeling or a theoretical enterprise conveys the notion that the relevant features and interpretation of rationality will be contemplated with methodological goals or criteria in mind. Hence there is a possible circularity in the understanding of rationality: we are looking for a correct understanding of it and the best theories and conceptual developments we can use to this purpose are themselves guided by methodological precepts which rely on ideas about rationality.

Relevance and accuracy in empirical explanation are valuable goals for model-building. However, explaining action as rational action inevitably involves some understanding of the motives or reasons for action, therefore some kind of interpretation of human conduct is needed. This contribution aims at investigating the main reasons why interpretation can hardly be severed from the explanation or understanding of rational decision-making. In spite of notable progress in the interactions of cognitive science and the social sciences, a markedly behaviorist posture has long been present in areas such as economics, in connection with epistemological empiricism and as a sequel to the grand methodological goal of developing a general model of decision-making (in contrast with the use of more or less ad hoc explanations in various domains). From this vantage point, interpretative operations might well seem redundant, given the all-important status given to the observation of choice operations. Nevertheless, there is a good case for having a second look at the relationships between the interpretation of action and decision-making models.

A working hypothesis, here, is that these relationships provide clues for the appropriate treatment of the normative components of rational action. Although there is a good case for seizing various opportunities to move ahead of the kind of rational choice modeling which is associated with traditional economic thought, it will be argued that the use of rationality assumptions is not automatically coupled with the disregard of those structural features of action which are exposed to open-ended interpretation. This will be exemplified by some of the rational-choice postulates which are commonly used. Moreover, getting a clear view of the interpretative tasks of social actors themselves helps understand the nature of social interactions, norm-based social mechanisms, and human action proper.

Decision-Making and the Interpretation of Actions in the Rational-Choice Perspective

The Economic Approach to Rationality as Model and Anti-model

Rational choice theory originates in philosophical developments about prudence and the securing of one's interest (with Aristotle, Hobbes and Hume as major contributors and sources of inspiration), and in a history of mathematical progress towards expected-utility theory (starting from Pascal's expected gain modelling and the classical model of Bernoulli—see Hacking (1975)—down to von Neumann-Morgenstern (1944, expected-utility theorem in the 2d. ed.) and Savage's (1954) personalist treatment). It has important links with the notion that social interaction makes it advisable to try to make full sense of expectations and meaning when it comes to interpreting the intentions, choices and actions of other human beings. Strategically speaking, this has been successfully exemplified in the works of Machiavelli (Barbut 1999). Contemporary discussions of the philosophical "principle of charity" (in the continuity of W.V.O. Quine and D. Davidson) also illustrate the permanent interest for the connection between rationality and the interpretation of human conduct.[2]

The theory is commonly associated in a dominant manner with the methodology of economics and, more particularly, with the use of optimization models in the field. It also reflects notions about consistency in choice which have major connections with common views about interests, passions and the need for justification; examples of well-developed theoretical explanations can be found in economics and economics-related fields (Picavet 1996). This makes it difficult to "use" the theory without conveying the notion that a more or less "economic" approach to human behavior is implied.

On the face of it, this is misguided, given the relevance of the theory in various fields. No matter how closely associated with economic methodology it may be, the range of rational choice theory exceeds the scope of economics proper. The understanding of human action it illustrates doesn't specifically rest on economic reasons in the usual and restrictive sense (relating to wealth or profit, or costs, or mutually beneficial exchanges, etc. with a view to self-interest, wealth or preference-satisfaction). Nevertheless, its special connection to economics cannot be overlooked; isn't it quite puzzling after all?

At the very least, this common association reflects equally common ways to identify a specifically "economic" dimension of reasoning and social life: reasoning as we do when we study economics often involves a focus on observable behavior and a methodological privilege for narrow pictures of antecedently defined interests on the part of human agents. Thus, an emphasis on rational-choice explanation is likely to be a vehicle for diminished expectations (with respect to realism and fine-grained accuracy) when it comes to interpreting human motives and perceptions in decision-making.

It remains true that the straightforward nature of explanations which are based on goal-oriented, narrowly motivated behavior in the economic sphere is a controversial conjecture. The role of interpretive tasks in this area of human behavior might be considered in its own right as it is in other domains of human life (the arguments below, in section "The Role of Assumptions About Preferences in the

---

[2]For an in-depth discussion which takes recent developments in cognitive psychology into account, see: Bonnay and Cozic (2011).

Development of an Instrumental Interpretation", point to this). First of all, it must be acknowledged that "narrow" motivations have an affinity with the framing of economic problems as specifically "economic", in contrast with problems whose analysis requires a more encompassing study of social life. In the economic order, to be sure, interest-driven abilities in decision-making play a conspicuous role. More often than not, they are presupposed in the rationale for given norms or institutions of this order. Moreover, although the goals of economic action can be plural, they are usually assumed to be amenable to description as a (composite) single objective, which is then treated as the *maximand* in an optimizing approach to the determination of action. This role can be played by the overall well-being derived from a variety of goods, or by the overall profit perspectives to be gained from multi-layered, often quite heterogeneous investments.

As a result, the main focus in economic-style rational choice modelling has to do with the adequacy between the choice of actions on the one hand, and goals or interests which are treated as "given". These goals or interests are often equated with the approximate psychological motivations (in line with Pareto's subjectivist turn) or, alternatively (and more objectively), with the predominant guiding rules for economic agents as they are (in their own society). The theoretician's ability to check the good correlation between the choice of actions and the best way to serve one's goals or interests is thus the key to a fairly general "economic approach of human behavior" [to use the phrase of Becker (1978)], with little attention paid to the phenomenology of the human way to entertain goals of a given kind.

Pursuing personal and corporate goals (or one's interests) turns out to be a focus for theoretical developments in a unifying perspective. The logic of such operations can be studied in an abstract manner, when generic goals such as "utility" and "profit" are under focus. Indeed, this is a topic for both economics and the general theory of decision. Inasmuch as the conclusions of such inquiries have a capacity to interfere with ordinary ways to deliberate about specific problems and values, they have a normative significance. It may happen that we trust the theory more than our ordinary ways of thinking and acting, and thinking before acting. This might be the case on account of the theory's systematic nature, interpersonal objectivity and unifying virtues with respect to the particulars of specific situations. A theoretician is then in a position to deliver pieces of advice of some significance, to the effect that our projected choices or behavior can be somehow corrected or improved.

The reasons which motivate our trust in theories are able to give credentials to such a correcting process. This is a strong point of the economic approach to human behavior, which stresses the unifying and systematizing virtues of a uniform approach to the very process of reaching a decision. In the process, our "corrected" decision-making operations will certainly exhibit biases and idiosyncrasies to a much lesser degree than otherwise, and looking at the conclusions of theories is a way to limit such imperfections.

The economic need for a simplified view of motivation [usually in association with a limited set of typical and important, mutually connected goals—as in Mill (1843), book 6] and the economist's concentration on optimization processes turns economics into some sort of applied, possibly prescriptive decision theory. To some extent, this should be qualified in view of the recent developments in experimental and behavioral economics, which makes it possible to investigate the specificities of our relationship to economic entities, such as monetary benefits, investment operations, wages etc. It remains true, however, that the bulk of economic analysis, in its general parts at least, is devoted to the examination of abstract, optimizing models of rational choice. Given the potential of such models for a "correction" of behavioral choice or effective action, this can be a source of criticism, turning economics into an anti-model from the point of view of our inquiries about the nature of human rationality.

Adjectival phrases which involve rationality, such as "instrumental rationality", "economic rationality", "optimizing rationality" are a testimony to this ambivalence. It does happen that they signal the high value to be credited to a clear-cut view of rationality, by virtue of which we become able to establish a clear correlation between values, beliefs and action. The same phrases might also signal, let it be acknowledged, the intrinsic limitations of a partial view of rationality as a human virtue or a practical disposition, with the implication that an ad libitum amplification of a "local" model of behavior, beyond the field of the study of economic affairs proper, is illegitimate after all, or at least ill-founded and somehow gratuitous. If economic behavior is described as a paradigm of rationality (Max Weber's *zweckrationalität*, or the choice of "logical" actions in Pareto's parlance), then we are at risk of falsely equating with rationality proper an economy-based, specific model of the involvement of reasons for action in deliberating about the actions to be taken.

Lessons from Criticism: The Limits of Generalization

The fairly impressive developments of the analysis of rational choice in economics provide a model of systematic explanation, argument and prediction in the social sciences. In fields like sociology, political science and management, the "economic model" is a source of inspiration for various schools of thought. More often than not, economic models which have been designed with a view to specifically explaining economic phenomena are being used or recycled in distant fields of research. This kind of theoretical generalization can be saved from caricature and, indeed, tentatively defended on methodological grounds. On the face of it, it is not illegitimate to use of model of the involvement of reasons in decision-making beyond the domain which has provided the main source of inspiration for its initial systematization.

The main reason is this: if the way economic reasons intermingle in decision-making makes sense in itself, independently of the economic significance of the goals under scrutiny, then an associated general model of reason-amalgamation might be relevant in other fields as well. Rather than an occasion for gratuitous generalization, this could be viewed with equal or superior insight as well-reasoned amplifying induction, because relevant knowledge is being put to use in domains in which it can shed some light. Nevertheless, there is no denying that the very process of a generalization, here, is vulnerable.

This appears to be associated, more often than not, with a reluctance to look at human behavior the economic way. Here, practical affairs and cognitive goals are associated, because economics can be used, as a system of arguments, methods and models, for the design of human systems of rule-based interaction. (accordingly, theoretical statements have a "performative" dimension which should be taken into account in economics, sociology, finance and ethics).[3] Apart from the design of rules, it can also be used as a source of inspiration for direct political or administrative action in given circumstances, thanks to its potential influence on contextual individual or committee judgment. It turns out that this is an object of potential rejection, given the consequences of resorting to economic categories, when the latter are being turned into a practical framework for deliberation.

An important objection is that such categories are a vehicle for a specific interpretation of actual and possible behavior (an interpretation which can be challenged). When human activities are collectively framed by an economic model, it does happen that self-oriented behavior is presumed, for example, even though the considered concrete activity might well testify to the effectiveness of motivations which take their origin in such concerns as one's faithfulness to antecedent or social commitments, obedience to rules (as an end in itself), or the goals of other agents (as an outcome of these agents' special prestige, superior know-how or distinguished social role for instance). In addition to self-goal behavior, economic models sometimes presuppose that other–oriented goals are excluded, so that behavior is to be considered self-interested, more or less after the pattern of egoism at the end of the day. Such developments are by no means necessary outcomes of the use of economic models to give a normative framework to human activities or transactions.

Given the prominence of simplified models with self-interested agents in models which can be used to frame or reform human activities (as evidenced in theory-based managerial schemes), economic rationality has somehow been associated with the reductionist hypotheses it often uses. Hence the rejection of so-called "economic rationality" as a framework for the conventional, socially embedded interpretation of behavior in given fields of social interaction. No matter how justified or ill-justified this claim is, let it be noted for our purposes that it doesn't really depend on the cognitive appropriateness of a modelling strategy. Rather, it relies on the practical appropriateness of an agreement on interpretative conventions.

A second argument we might consider in order to mitigate the diagnostic of illegitimate generalization is that economic affairs haven't been the only source of inspiration for the elaboration of an economic model of behavior. The influence of the rhetorical justification of action (Aristotle), military analytics (Machiavel, Clausewitz), political reasoning (Hobbes) and the mathematical study of betting and games (from Pascal through Bernouilli to the present) have been conspicuous. This circumstance is a reminder that the impact of so-called "economic" reasons in

---

[3]See Brisset (2014), MacKenzie and Millo (2003), Walter (2012).

decision-making operations is perhaps not typically "economic" in the first place, hence the arguable legitimacy of an heuristic strategy in which existing, well-developed economic-style models of rational choice are used a benchmarks outside the field of economics in a narrow sense.

An intermediate, interesting possibility materializes when the "economic" layer of rationality is understood as a kind of rationality which is of special significance for economic behavior in a narrow sense (the kind of behavior which relates to unmistakably economic evaluations, deliberations and transactions). It is conceivable that this kind of rationality is of some significance in other fields as well; this can be investigated, using "economic" models in an exploratory way, as a means to identify and systematize individual motivations and social equilibria in a given domain of social interaction. To be fair, this has not really proved fully convincing up to now[4] and it might well be the case that, after all, economic rationality is well-suited for inquiry into a limited number of economic problems, and nothing more. Using economic models as a model of rationality would then involve an interpretative choice which is based on analogy rather than realist modelling.

The plausibility of relying on economic rationality for non-economic problems is sometimes advocated by referring to the intrinsic importance of the economic dimension of a correct use and allocation of resources. For example, ethical guidelines about medical acts and organizations sometimes give advice about benefit-cost effectiveness, in the economic fashion. More controversial references to economic reasoning originate in the belief that specific pragmatic and social problems predominantly reflect the constraints to be derived from limited resources, egoistic goals, and a strategic setting of actions (including the existing mutual threats and the possibilities for alliance-building). The study of such problems is then trusted to game theory and microeconomic-style models. Given the limited resources, the assumed antecedence of goal-setting and the strategic nature of interaction, a rather limitative view of human motivation might appear sufficient for explanatory purposes.

This can be taken as a testimony of the dual nature of concerns about realism. On the one hand, they invite researchers to look after detailed, convincing accounts of motivation or reasons. On the other hand, when behavior turns out to be severely constrained in its possibilities by strategic calculus, and framed by antecedent determination on the volitional side, then it might seem "realistic" not to be influenced by psychological particulars, focusing attention inside on a core set of major, objective parameters. This can be further defended on positivist grounds, starting from ideas about the correlation of observable events and scientific language [see Caldwell (1982)].

A further possibility to be considered is that the kind of rational-choice model which has been devised to advance the understanding of economic affairs has undergone a process of excessive insulation from either empirical validation

---

[4]See, for example: Boudon (1998), Green and Shapiro (1994), Plon (1976). A more optimistic account can be found in: Attali (1972), Becker (1978).

(Demeulenaere 1996), common-sense evaluation (Searle 1995), or reflective practical reasoning, to the effect that its ability to be relevant in its native economic domain is at best doubtful in the first place. A recommended strategy, then, is the gradual enrichment of the economic models of rational choice, looking for a better fit with key insights about ordinary human behavior. Looking after the complementarities in various fields of study, or theoretical disciplines, is then advisable as an alternative to the generalization of economic benchmarks. It is then a consistent option to develop a consciously interdisciplinary approach to behavior and human reasons, looking for a general approach to reasons and motives [see for example Bouvier (1999)]. Weber's *homo oeconomicus* type should not be viewed, then, as a human agent whose reasoning abilities are so severely restricted that he can only act like automata performing a role in the simplest of economic models.

Common Sense and Theoretical Interpretation in the Framing of Action

Human actions raise questions which relate to their interpretation: which meaning should we give them in the light of concerns which are attributed to the actor? How are they perceived, and sometimes misperceived, by other social actors? And also: should we give them meaning only in light of those facts which are confined within the sphere of data accessible to the player himself? These are questions for both ordinary life and scientific methodology. Maybe we should believe, with Fénelon, that common language should not prevail over the true principles.[5] When man and society are the object of study, however, the relationships between social experience and theory can be of some help to understand the real benefits of theory as well as its limits.

Our actions are objects of interpretation and it seems impossible to understand them, except with a correct understanding of this interpretation process as it takes place in social life. Presumably, this realist requirement is a minimal one and it is grounded in familiar facts about human action. Social scientists, however, are theoretically minded in their explanatory efforts, and popular conceptions as well as ordinary language and the ad hoc penetration of motives in singular cases are possible hindrances to systematic explanation. Making use of systematic decision analysis, in particular, involves a number of departures from the everyday treatment of action in ordinary life and discourse.

Thus, what one would describe as inaction in ordinary conversation must be re-described in terms of positive choice if we interpret action as chosen out of good reasons; insofar as *other* choices are possible, the perception of an attitude as "inaction" doesn't nullify the fact that a choice is being made, and this choice is

---

[5]Fénelon, *La Nature de l'homme expliquée par les simples notions de l'être en général.* In *Oeuvres,* vol. 2, Paris, Gallimard, «La Pléiade» series, p. 851

meaningful in its own right. In addition, the "consequences" which are considered in decision models jointly result from the action of the agent, natural events as well as the actions of others (provided the agent is not alone). Generally speaking, the so-called "consequences" in decision models can hardly be identified with the "consequences of action" in a narrow sense, that is to say, the specific outcomes of undertaken actions, which can be imputed to the agent with a degree of exclusivity (once a breakdown of the intentional and non-intentional parts of causal impacts in the circumstances has been accepted).

Considered from the perspective of decision theory, human action is not truly defined regardless of the consequences. Action and consequences are typically considered together and in a correlated manner, in sharp contrast with the world of everyday human activity. In this "real" world, statements which describe the consequences are typically formulated independently of the statements which describe how these results can be achieved (and how they are in fact achieved). In decision theory and in the economic and social models which are dependent upon it, the actions must, of necessity, be correlated with the possible states of the world, in such a way that they qualify as the available means which can be used in order to achieve the desired results. What amounts to a requirement of good theory-building turns out having substantial consequences for the description, interpretation and explanation of action.

Thus, the rational-choice view of human action is elaborated in such a way that the choice of an action can be identified with a goal-oriented striving after results. Looking for the counterpart in common-sense descriptions of ordinary action in everyday life would be quite an intriguing exercise. Thus, what about describing the result to be obtained in a way that depends on how one describes the proposed means in the first place? Maybe this happens in some cases, but this is by no means a usual linguistic attitude. One may think that, should it happen, it would really amount to a simplifying way to speak, avoiding heavier, more precise descriptions which remain possible in principle. For example, if I say that I intend to go to the nearest possible pharmacy, it is true that the result which is implicitly targeted (getting drugs from a chemist without losing too much time) apparently depends on the action that leads to it (choosing the shortest route, according to a certain strategy of exploration of the territory. In addition, the action which I describe seems to be somehow goal-oriented: it really amounts to seeing to it that the result obtains. However, I am aware that I can describe it another way, perhaps in a slightly more complicated manner, describing the series of plans and physical operations and choices through which I intend to pay a visit to the chemist. In ordinary discourse, I surely won't, yet it is possible in principle at least.

In the light of decision theory, an action is precisely the correspondence established between external (exogenous) events and the possible consequences which can be enumerated; it is nothing else at the theoretical level: actions are just ways to determine the consequences of possible events. When it comes to interpreting the theory, the well-defined actions are then usually identified with actions in the ordinary sense (various initiatives of the agents followed by gestures of these agents in the physical world, usually with a degree of adjustment to the actions of other people). Insofar as the choice rule is described in the form of optimizing an objective function, it is customary to describe it in instrumental terms: choosing an action with an "optimizing" property would amount to reaching the goal with maximal success, thus making action the equivalent of a good instrument. In reality, one must consider the desirable and less desirable qualities of the consequences, and the implications of uncertainty or risk generally speaking.

As soon as uncertainty is brought in, a certain distance is established between object selection and the ultimate object of preference. Indeed, it is then no longer immediately possible to interpret one's choice as the instrument for achieving what is best. The outcome might well be the worst of outcomes, even if the agent is rational. In other words, taking an aggregate view of the possible consequences of each possible action is required. If this view is a balanced and well-reasoned one and if a correct course of action follows, then the action can be considered rational. It is perfectly possible, however, that it leads to very poor results in the end. Thus, choice selection based on expected utility often squares more or less with common views about good reasons for action in the face of risk [with the usual qualifications which are made necessary by such paradoxes as Allais's (1953) and Ellsberg's (1961)] but allowing this cannot lead us to forget that expected utility maximization doesn't secure good results. This perspective on action can be contrasted with common uses of the intention-result pair. In ordinary language, unexpected results, especially the bad ones, aren't just possibilities among others; they are also framed as unintentional results, that is to say, results that somehow shouldn't have obtained (given the monitoring of action in the light of a certain intention).

Following classical decision theory (the axiomatics of Leonard Savage (1954) remain emblematic), actions are not in themselves the ultimate source of motivation, even though they are the proper object of choice. The actions appear desirable or not, given a certain perspective on the profile of their possible consequences. To understand the action - to explain it and to have a grasp of the issues as they occur to the agent - you have to know how the agent considers the possible consequences and their action-related probabilities. This, again, is in sharp contrast with ordinary views of action-explanation. In everyday life, there is little doubt that we are prepared to accept explanations of actions which follow from a "will" to act in a given way (suggesting that the ultimate reasons have to do with the action itself) or, alternatively, from an "intention" to get a certain kind of results (with the suggestion that the other possible results are anomalies). The implied picture of action is decidedly different from the theoretical one.

Dealing with choice in situations of risk or uncertainty, decision theory tries to develop an intuition of consistency in the practical attitude, in the face of possible states of world, focusing on the values and beliefs of the agent. The conditions that we impose on preferences make some kind of consistent personal attitude explicit; they synthesize relevant reasons or judgments (the criteria of choice). We study the perspective of the actor on the possible outcomes of conceivable actions, even if we admit that there is a known distance between mental content and explanatory hypotheses (Simon 1959).

Values, Interpretation and Interactions

Identifying Goals or Values: The Interpretative Dimension

Identifying the values or goals of real-world agents is no easy task. Taking such motives as "given", however, makes it possible to refer to them as hypothetical components of the explanation of behavior. More often than not, this turns out to reduce the apparent role of interpretive operations, while cost-benefit calculations and deductive reasoning (based on optimization models) ostensibly take the lead.

Even in economics though, this theoretical outlook has its problems. The exact nature of "optimization" in "optimizing behavior" calls for attention. At the core of economic reasoning as it is usually described, we find the notion that human behavior should make the best of existing resources, at the individual level (for a given social structure) and from a collective perspective as well. But what does a phrase like "the best" stand for in the first place? To give content to it, should we refer to expected utility theory or, given the experimental refutations, to one of the rival theories (and which one)? Is "optimizing" itself a quick way to refer to general guidelines of behavior which originate in the constraints of the social, natural and technological environment? Or should we rather view it as a general model of the mental operations involved in decision-making tasks? The latter view isn't really congenial to the main currents in traditional economic theory but, given that economic theory increasingly uses the results of experimental psychology, the relevance of such a view cannot be ignored.

Whether they are taken to describe the actual mental processes in deliberation or the general orientation of action-taking, economic models of optimal choice usually reflect the theorist's prior choice of a set of interests (or preferences) which are conspicuous in the problems at hand. This is illustrated by the self-conscious choice of a limited number of influential factors and motives, much in line with the methodological advice for the practice of economic explanation in J.S. Mill's *System of Logic*. The achieved simplicity in the ultimate determinants of choice behavior is thus, by and large, the result of a chosen methodological outlook.

Consequently, the observed choices can be equated with vehicles of the pursuit of the selected interests or values. With this in mind, simplifying hypotheses prepare the ground for warranted simple interpretations of behavior. Complexities are present to be sure. Ultimately, however, they do not relate to complex perceptions of action itself, at least in the simpler sort of classical theories. Rather, complexities typically have to do with finding out the best way to go ahead in the pursuit of one's interests or values, given the specificities of the situations under scrutiny. Awareness to complexity follows from the attention to such key features as the sharing of information, the presence of uncertainty or risk, and the institutional spectrum of possible social arrangements in the existing legal framework.

In various fields of theoretical explanation beyond economics, such as political analysis and various branches of sociology, choice selection procedures and the social interactions to be considered are quite complex after all. They turn out to depend on language, historical circumstances and cultural factors that give meaning to human action itself. Often, no self-obvious goals or interests can be taken for granted, which amounts to saying (in Herbert Simon's categories) that a "structuring" phase of decision-making activities must be taken into account, so that one gets a clear (or at least a clearer) mind about what is at stake in decision-making tasks (for individuals or groups).

In such circumstances, the necessity of developing interpretations of intentional action and its underlying reasons, in light of a wide array of possible reasons, is even clearer than in economics at first glance. These reservations being done, the rational-choice hypothesis can still play a very important role in positive theory as exemplified by political analysis, (for predictive and explanatory purposes, e.g. in the spatial model of voting, or in applications of game theory to the field of international relations) as well as within normative or prescriptive social and political theory (important impacted branches include contractarian approaches to power or social justice, international strategy and study of rational negotiation, analyses of acceptable or reasonable compromise schemes, collective choice theory). One of the attractive features of rational choice theory is the inferential versatility and fertility of optimization models.

For these merits to materialize, though, these models should account for the main known tendencies in behavior in an effective way. Using rational choice theory in the process of explaining human action in complex environments, without relying on drastically simplifying assumptions about the interests to be pursued, we can hardly place our bets on the ability of the pragmatic environment to generate constraints on behavior which are sufficient for an actual optimizing process to take place some way or other. In such complex settings, are we certain that individual agents are able to process information and monitor decision-making in such a way that the resulting choices and actions are of an "optimizing" sort? Following Herbert Simon on the so-called "bounded rationality" path, many researchers opt for a negative answer.

Incidentally, means-end reasoning is an often misleading way to synthesize the process of an optimal choice of actions. The usually composite nature of the values or interests endangers the meaningfulness of means-end parlance. When avoiding costs is involved, not just achieving positive results, is it really the case that decision-making is governed by (positive) goal-seeking? When several individuals are considered, what is the point of considering "welfare" or "well-being" as a goal? Shouldn't we rather acknowledge that a plurality of goals is involved in the social process and in the social patterns of individual actions?

These questions are very pressing when an optimal-choice modelling strategy is applied to areas (social or political areas in particular) in which the structure of choice and interaction is obviously and quite heavily dependent on language and cultural interpretations. The social processes of interpretation can hardly be ignored, then: they are essential ingredients of the available choices. A relevant choice is a choice which can be validated as a credible interpretation of what it means to act in such and such way in a given class of circumstances; These circumstances are themselves interpreted as examples of relevant benchmark types of situations, which call for given reactions (in accordance with such guidelines as principles of morality of prudential rationality, religious doctrines, lifestyles or the shared expectations in a social milieu—this is not meant as a closed enumeration).

As a result, it is hardly possible to take for granted that existing, pre-determined goals are being pursued. Such a simplifying strategy has its plausibility in given, limited domains of choice (such as given categories of economic or technological choices). It can hardly be generalized as a universally valid strategy. The next epistemological question to be handled, then, is the legitimacy of a separate treatment of a set of motives in relation with a separate group of choice situations. We will not review here the existing skeptical arguments about this, which reflect concerns with the realistic nature of the account of intentional choice and its motivational structure.

Even in the case of economic-style concentration on a very limited set of interests at play, the very process through which such a set of interrelated interests is selected is connected with interpretative operations: some areas of choice behavior are interpreted as interest-driven, for a given understanding of the relevant interests. In the general case, intermediate goals are being pursued at successive stages in decision-making processes, within a larger frame in which evolving beliefs, interests, values and principles are guiding forces for the social agents. The associated interpretative operations cannot, then, be overlooked in the explanation of choice behavior.

Interpretation of Choice and Optimizing Models

Given the discrepancies between theoretical and practical languages about action, should we come to believe that the conventional postulates of rational choice, whose explanatory potential is well documented in economics, are best understood as short ways to the interpretation of action? In many respects, this seems plausible at first sight, because optimizing models, with their elaborated relationships to the axiomatics of rational choice, tend to suggest definite interpretations of action: making the best of the circumstances, acting in the best possible way, consistently fulfilling one's desires in light of one's beliefs. This suggestion, however, requires some elaboration. The old notion of choosing the best or the greatest good (which has long played a founding role in many systematic normative theories, such as the political theory of Hobbes) isn't always clearly related to the analytics of action. Decision theory provides the analytics, but how can it build a connection with "choosing the best"?

The list of conditions imposed on choice in decision theory typically includes conditions such as transitivity, the absence of "cycles" of preferences, Sen's *alpha* and *beta* conditions. If the choices of an agent are supposed to obey a scheme of this kind, the observed choices will easily be interpreted as a consistent way to follow one's preferences. The foundations of the analysis here amount to a postulated structure and the conditions imposed on this structure so that an interpretation of choice as the consequent election of the best is made possible.

This applies to individual economic agents but also to institutional players, which are not endowed with the psychology of human agents but whose choices are understandable or explicable by reference to "acceptance" (instead of beliefs) and collective "values" (instead of individual desires) as revealed in the transactions of institutional life. The interpretation is crucial and if we do not stick to a strictly behavioral approach, it can be backed by verbal documents: public statements, doctrinal productions of institutions, etc.

There is indeed in the interpretation of choices an operation analogous to that of the application of the "principle of charity" in discursive exchange: if I decide to interpret the statements of others as rational statements, I choose to identify a certain set of linguistic productions with a theoretical object, which can be described in the following way: what would be said by a reasonable individual with a certain number of beliefs, accepting certain rules of inference and aiming at certain goals in the circumstances, and given the structural traits of the communication situation. Clearly enough, such an object is by no means immediately given in empirical data; It is constructed by the analyst through a process which has a dimension of normativity because the "reasonable" is under focus; then it is substituted for the observable phenomenon (the issued statements) in order to explain it. Similarly, decision theory establishes the correspondence between an empirical object (observable choices), a theoretical object (the choice made consistently on the basis of consistent - for example complete and transitive—preferences) and an explanatory strategy (the reconstruction of choices as products of an optimization process in the light of an objective).

It should be noted that the existence (guaranteed under certain assumptions) of a function whose optimization coincides with a non-dominated choice (given a preference ordering) does not mean that the agent really aims at a goal expressed by this function. It is not implied that his or her mental state would be to target that goal or the means by which it could be achieved. Furthermore, it might be the case that several distinct functions could do the job: several descriptions of objectives whose implications for choice would be the same. The problem is similar in substance to the problems of under-determined interpretation, as studied in the context of debates on the charity principle.

The procedure for assigning an objective function to concrete agents, for example the "utility function" of economists, envelops a procedure for interpreting observable choices. This interpretation is important, from an epistemological point of view, for the economic and social sciences. Indeed, in most families of models of behavior, especially in relation to the human species, an attempt is made to explain the behavior of the agents on the basis of a sort of understandable compromise between some essential motives, which are identifiable (e.g. leisure and money in a theory of labor supply, the advantages of conquest and the disadvantages of losses in the case of military actions, etc.). This compromise being relatively simple in most cases, it might look like a composite goal.

Insofar as one seeks to explain and predict the agent's conduct in this way, it is understandable that one gives oneself a theoretical object which has the structure associated with the intended form of the explanation (a rational agent, a function which is more or less similar to a goal). However, it can be observed that the interpretation of the observed behavior in terms of optimization of the objective is legitimate only under a very strong hypothesis, according to which choices are precisely the selection of the preferable (that this selection must be assumed "voluntary" or not is yet another problem).

The kind of "instrumental" interpretation that is made possible by the conditions imposed on preferences is thus actually a structure of interpretation of the action, rather than a structure of the action itself. Indeed, there is a necessary link between the interpretation of action as an instrumentally correct action and the interpretation of action as an initiative oriented towards the preferable. We can see this in the following way, reasoning in the case of «certainty» (i.e. in the absence of uncertainty). The objective of an instrumental action can only be the obtaining of the preferable: it is indeed because the action is supposed to favor the attainment of a result deemed preferable that the action appears as an instrument for the actor. Indeed, one uses an instrument only because one aims at something and what the non-schizophrenic agent aims at is identical with what he or she prefers.

Action can be treated as instrumental (in the formal sense of the possibility of a representation in terms of optimization of an objective function or saturation of a structure of preferences) because preferences are supposed to exhibit features which warrant that the preferable has a definite meaning, thanks to the postulates of coherence (transitivity, quasi-transitivity or acyclicity) and completeness (the comparability of any pair of possible choices). Thus, the search for the preferable and the instrumental character lent to the action go hand in hand. These are two correlated aspects of an action-explanation scheme. This scheme favors the reasons of the actors, not just a formal structure of choices. The reasons in question relate to certain ways of articulating values on the one hand, and beliefs (or acceptations) on the other hand.

The representation of action which is conveyed by the theory of decision isn't strictly instrumental in character at first glance, even though it must illustrate the tautology that the preferred choice is what one should choose if one is to act in order to satisfy one's preferences. Once a specified objective is under focus (optimization of a function-objective or saturation of a structure of preferences), an object of study is proposed, about which it is meaningful to ask the following question: do certain possible choices express the pursuit of the specified objective in a consistent way? The emergence of such an object of study occurs thanks to the endorsement of hypotheses about the structure of preferences, oriented towards a certain interpretation of human action - namely, an "instrumental" interpretation concerning goals or a composite system of ends which are reconstituted for scientific purposes.

The Role of Assumptions About Preferences
in the Development of an Instrumental Interpretation

We can appreciate quite simply the importance of assumptions about preferences in the design of a goal-oriented behavioral model. Let us suppose that we have a rough explanatory basis, consisting of the preferences of the agent. Let us put ourselves in a situation where there is no uncertainty: the choice of an option is the choice of a state of the world.

Traditional decision theory requires that an agent is indifferent between a choice and the same choice. Indeed, if it were possible, for a certain option a, that a were strictly preferred to a, then there would be a problem with specifying the preference structure. The interpretation of choice in the light of individual preferences, as a "choice of the preferable", would be endangered. If the results of a questionnaire tend to accredit preferences of this kind, it certainly appears necessary to try to redefine the objects of choice so that the choice of a can be interpreted as the choice of an option which is better than other options, but not better than itself. To do this, one would have to ask in particular whether, when an individual says that he or she strictly prefers a to a, he or she does not actually consider distinct objects under label a.

If we did not suppose completeness, and if the choice of a came to be observed, b being available as an alternative choice, it could be impossible, owing to the lack of completeness, to allow that the choice of a allows the agent to better satisfy his preferences than he could have done by choosing b. Indeed, the two objects may be incomparable, given the data about preferences. We could not then interpret the exclusion of b as the result of a preferential choice from the set of available options. Even if the available model of choice was able correctly to predict the choice of a when a and b are available, one could not go so far in interpretation without arbitrariness.

Let us suppose that we do not postulate the transitivity of preferences (nor related, weaker conditions) and assume that we observe the following choices:

(1) the choice of a while a and b are available,
(2) the choice of b while b and c are available,

Now suppose that on the basis of an instrumental interpretation of the choices it is inferred from (1) that a is chosen because it satisfies the objective better than b. It is similarly inferred from (2) that b is chosen because it satisfies the objective better than c. From the instrumental point of view, it would be necessary to admit that a enables the agent to better achieve its objective (to attain it to a greater degree, to get closer to it, etc.) than c.

However, it cannot be excluded that the agent prefers c to a, although a achieves the presumed objective better than c (being admitted that this presumed objective is a goal worthy of the name, respecting transitivity). Transitivity of preferences is a postulate that allows aligning the preferences and a classification of the choices according to their instrumental value. Indeed, the preferences assumed in choices can, under this postulate, be systematically matched with the instrumental value of the choices—"preferred to" being aligned with "going more in the direction of the objective aimed at".

Thus, the hypotheses about preferences make it possible to interpret the observation of choices as preferential choices being made in the pursuit of a (possibly composite) objective. If we rely on the hypotheses on the preferences which have been mentioned, we can propose an instrumental model of optimizing behavior. The fact that it is possible and appropriate to explain the observed conduct in terms of optimization of a certain objective function is a reason to accept the identification of the object of study to a behavior faithfully reflecting the selection of what is «the best» in the light of consistent value judgments (individual preferences).

Values and Interpretation

In the study of institutional interactions it is difficult to leave aside the norms that are laid down, which serve as a reference for public debates and institutions and which may have an impact on the social processes through complex modalities of argument and deliberation. Thus, in spite of the very refined nature of the behavioral foundations of economic analysis, empirical studies of social life—and political economy itself—deal with norms and institutions, considered both in themselves and through their constitutive value for the interpretation of initiatives or reactions. On the face of it, this follows in a straightforward manner from the insertion of individual and group action in organized social life.

This raises the problem of the explicit consideration, in the explanations of normative and institutional dynamics, of the very meaning of norms. Indeed, the influence of institutionalized norms is not only due to their capacity to support incentive mechanisms (benefits, praise and trust in cases of respect, formal or less formal sanctions in case of violation). It is also linked to their contribution to the interpretation of each other's actions. For example, the violation of a norm that no one can ignore, in a way that cannot escape attention, can be perceived as the expression of a normative claim rather than a mere attempt to reap the fruits of norm violation.

As a first approximation, the meaning of the norms is fixed by their criteria of correspondence with situations (which are said to be compliant or non-compliant on the basis of these criteria) and by the use of these standards in the qualification of the situations themselves (statements which provide descriptive elements for the identification of situations to be relevantly classified as compliant or non-compliant). However, these verification and qualification criteria may evolve considerably under the influence of the expertise developed by various groups in society or by the institutions themselves in some cases.

In addition, through descriptive typologies and models of policy implementation procedures (with or without conflict between agents), we have gained insights about the forms of governance that are structured by partly indeterminate norms (Matland

1995; Jones and Clark 2001; Reynaud 2003; Picavet 2006; 2013). Progress has been made in general representations of negotiation and rule-based concessions [see Kersten et al. (1988)]. In these different approaches, the interpretation of norms plays a role as an ingredient of social interaction.

The model of the political and political-economic organization that remains dominant in common representations is that of concerted action around common objectives. But when institutionalized norms play a decisive role in the perception of the actions among the social actors themselves, we cannot ignore the impact of these norms—and their most influential interpretations—on the regime of interactions between centers of power. We must then look at the interpretative conventions which account for the accepted use of powers and the state of real prerogatives (beyond the formal power resulting from a nominal reading of normative statements). These conventions result, among other factors, from the accepted initiatives on the part of the power centers (whose authority itself reciprocally results from the system of norms).

The concrete modalities of public decision-making are then determined through the interpretative and strategic relationship to norms. For example, in the strategic interaction between Parliament, Commission and Court of Justice in the European episode of gas emission, following Moser's (1997) analysis a pivotal role is played by the interpretation of ecological demands in the light of opening up markets in the Union. Other examples can be found in the strategic interaction between the national executive, the Commission and courts around the area of economic competence of the State and the legality of direct interventions in the productive sectors (Le Galès 2001; Picavet 2006).

In such examples, the institutional actors appear to be influenced by estimates of the plausibility of possible interpretations of the norms of the European Treaties, and also by representations concerning the normal forms of economic activity on the part of the State (direct economic action, imperative regulation, flexible regulation). In these examples, some norms seem to be part of the framework of the preferences which are manifested in the choices which occur in institutionalized procedures. Let us say, more generally, that they play a constitutive role in the reasons of the chosen strategies.

The interpretation of legitimacy norms thus appears important for economic life. To the extent that the preferences and beliefs of certain agents relate to the distribution of powers among the decision-making centers, making norms effective depends on the endogenous evolution of beliefs and preferences about the correct distribution of effective rights, powers or prerogatives. From an epistemological point of view, it should be stressed that in the developments of the politico-economic theory of the institutions, there is a strong link between "positive" and "normative" aspects, with issues of rationality as a common ground.

We may wonder how interactions are structured by generally accepted norms and principles. One of the general questions in the background can be formulated as follows: what can we learn about the nature of action according to principles, or the implementation of principles, in a world of institutions? In order to make progress in this direction, positive inquiry must develop connections with social ontology (about the nature of action and power). Moreover, the results of ontological questioning certainly matter if one intends to reflect on the desirable organization of interaction, communication and deliberation.

Conclusion

The fundamentals of rational choice theory do not commit us to a restrictive methodological view which would reduce the explainable features of human behavior to the pursuit or narrow and specific goals, or pre-determined patterns of usual (and therefore expected) behavior. They are to be found, rather, in concerns about consistency in choice (and in valuing operations), in well-ordered preferences about alternative states of the world, and in the consistency of a perspective on possible events, or the adequacy of beliefs generally speaking.

Letting hermeneutics enter the picture doesn't equate to a retreat from rational-choice explaining strategies in methodological individualism. The discussion of interpretation highlights the relevance of rational-choice thinking to some degree. Indeed, the legitimate quest for a unified and general explanatory model of decision-making—a substantial motivation of researchers when they opt for rational-choice methodology—cannot provide support to an extreme reductionist posture which only allows for the observation of external behavior, with no attempt at interpreting the norms and substantial reasons which influence or determine our actions. This should be taken as good news for methodological individualism after all, for all the undisputed impact of behaviorism on the elaboration of modern microeconomics.

Upon examination, rational-choice postulates illustrate the need for a self-conscious interpretation of human action. Accordingly, theories of rational decision-making must provide a room for interpretation because it is fundamental to their grasp of actions as intentional actions in the general view of human action they take. Moreover, interpretive tasks lie at the heart of social interaction and they are both framed and impacted by the institutions. For this reason, interpretation cannot be reduced to a methodological choice or a posture of understanding that would be projected onto the social. In many respects, it is part of the fabric of economic, institutional and social life. Some way or other, the explanatory theories can only reflect this embeddedness of interpretation in human interaction.

References

Allais, M. (1953). Le comportement de l'homme rationnel devant le risque: critique des postulats et axiomes de l'Ecole américaine. *Econometrica, 21,* 503–546.
Attali, J. (1972). *Analyse économique de la vie politique* (2nd ed.). Paris: Presses Universitaires de France (1981).

Barbut, M. (1999). Machiavel et la praxéologie mathématique. *Mathématiques, informatique et sciences humaines, 37*(146), 19–30.

Becker, G. (1978). *The economic approach to human behavior.* Chicago: The University of Chicago press.

Bonnay, D., & Cozic, M. (2011) Principe de charité et sciences de l'homme. In T. Martin (Ed.), *Les sciences humaines sont-elles des sciences?* (pp. 119–158). Paris: Vuibert.

Boudon, R. (1998). Au-delà du 'modèle du choix rationnel'. In B. Saint-Sernin, E. Picavet, R. Fillieule, & P. Demeulenaere (Eds.), *Les Modèles de l'action* (pp. 21–49). Paris: Presses Universitaires de France.

Bouvier, A. (1999). *Philosophie des sciences sociales.* Paris: Presses universitaires de France.

Brisset, N. (2014). *Performativité des énoncés de la théorie économique: une approche conventionnaliste (PhD thesis).* Université de Lausanne and Université Paris 1 Panthéon-Sorbonne.

Caldwell, B. (1982). *Beyond positivism: Economic methodology in the twentieth century.* Londres: Allen & Unwin.

Calvert, R., & Johnson, J. (1998). Rational actors, political argument, and democratic deliberation. In *Annual Meeting of the American Political Science Association.* Manuscrit disponible (version 1.2) sur internet, site de l'Université de Rochester.

Demeulenaere, P. (1996). *Homo oeconomicus. Enquête sur la constitution d'un paradigme.* Paris: Presses Universitaires de France.

Demeulenaere, P. (2003). *Les normes sociales entre accords et désaccords.* Paris: Presses universitaires de France.

Ellsberg, D. (1961). Risk, ambiguity and the Savage axioms. *Quarterly Journal of Economics, 75* (4), 643–666.

Green, D. P., & Shapiro, I. (1994). *Pathologies of rational choice theory. A critique of applications in political science.* New Haven: Yale University Press.

Hacking, I. (1975). *The emergence of probability.* Cambridge: Cambridge University Press.

Jones, A., & Clark, J. (2001). *The modalities of European Union Governance. New institutionalist explanations of agri-environmental policy.* Oxford: Oxford University Press.

Kersten, G., Michalowski, W., Matwin, S., & Szpakowicz, S. (1988). Representing the negotiation process with a rule-based formalism. *Theory and Decision, 25*(3), 225–257.

Le Gales, P. (2001). *Est Maître Des Lieux Celui Qui Les Organise*: When National and European policy domains collide. In Stone Sweet (Alec), Sandholtz (Wayne) et Fligstein (Neil), dir., *The Institutionalization of Europe* (pp. 137–154). Oxford: Oxford University Press.

Mackenzie, D., & Millo, Y. (2003). Constructing a market, performing theory: The historical sociology of a financial derivatives exchange. *American Journal of Sociology, 109,* 107–145.

Matland, R. E. (1995). Synthesizing the implementation literature: The Ambiguity-conflict model of policy implementation. *Journal of Public Administration Research and Theory, 5*(2), 145–175.

Mill, J. S. (1988). *A system of logic, ratiocinative and inductive: Being a connected view of the principles of evidence and the methods of scientific investigation* (1843, 6th éd. 1865), Book 6: *The logic of the moral sciences.* Peru, Illinois: Open Court.

Moser, P. (1997). A theory of the conditional influence of the European Parliament in the cooperation procedure. *Public Choice, 91,* 333–350.

Picavet, E. (1996). *Choix rationnel et vie publique.* Paris: Presses universitaires de France.

Picavet, E. (2006). L'institutionnalisation de l'attribution des pouvoirs politico-économiques: normalité et exception. *Revue Canadienne Droit et Société/ Canadian Journal of Law and Society, 21*(1), 39–62.

Picavet, E. (2013). Neoliberalism and authority relationships. In Merle, J. -C. (ed.), *Spheres of global justice* (vol. 2). Springer: Dordrecht (Chap. 52).

Plon, M. (1976). *La Théorie des jeux: une politique imaginaire.* Paris: Maspéro.

Reynaud, B. (2003). *Operating rules in organizations. Macroeconomic and microeconomic analyses.* London: Palgrave.

Savage, L. (1954). *The foundations of statistics*. New York: Wiley. (2nd ed. 1972, New York, Dover).

Searle, J. (1995). *The construction of social reality*. New York: The Free Press.

Simon, H. (1959). Theories of decision-making in economics and behavioural science. *American Economic Review, 49*(3), 253–283.

Von Neumann, J., & Morgenstern, O. (1944). *Theory of games and economic behavior, princeton* (2nd ed., 1947, 3rd ed. 1953). NJ: Princeton University Press.

Walter, C. (2012). Ethique et finance: la tournant performatif. *Transversalités, 4*(124), 29–42.

CPSIA information can be obtained
at www.ICGtesting.com
Printed in the USA
LVOW13*1942050818

586024LV00004B/19/P

9 783319 940267